Back to Basics in Physiology

Back to Basics in Physiology
Fluids in the Renal and Cardiovascular Systems

Juan Pablo Arroyo
Adam J. Schweickert

ELSEVIER

AMSTERDAM • BOSTON • HEIDELBERG • LONDON
NEW YORK • OXFORD • PARIS • SAN DIEGO
SAN FRANCISCO • SINGAPORE • SYDNEY • TOKYO
Academic Press is an imprint of Elsevier

Academic Press is an imprint of Elsevier
The Boulevard, Langford Lane, Kidlington, Oxford, OX5 1GB, UK
25 Wyman Street, Waltham, MA 02451, USA

First published 2013

British Library Cataloguing in Publication Data
A catalogue record for this book is available from the British Library

Library of Congress Cataloging-in-Publication Data
A catalog record for this book is available from the Library of Congress

ISBN: 978-0-12-407168-1

For information on all Academic Press publications
visit our website at store.elsevier.com

This book has been manufactured using Print On Demand technology. Each copy is produced to order
and is limited to black ink. The online version of this book will show color figures where appropriate.

Working together
to grow libraries in
developing countries

www.elsevier.com • www.bookaid.org

CONTENTS

Perhaps one of the most difficult transitions after graduating medical, physician assistant, or nurse-practitioner school and entering residency or a hospital based-care environment regardless of discipline, is integrating what one has learned about discrete organ systems with systemic response to illness, injury, and therapeutic intervention. The complex interplay of diverse organ systems is commonly mediated through the circulatory system, regardless of whether that link is driven by oxygen content, mean arterial pressure, or carriage of hormonal mediators. Understanding the fundamental underpinnings of fluid mechanics and dynamics with and without cellular elements and oxygen helps the clinician span the gap between knowledge and integrated practice. This text is straightforward, simple, and direct spanning the basics of fluid distribution through oxygen transport, renal function, and systemic fluid handling including clinically relevant vignettes one might encounter in daily practice. These essential concepts are taught using plain language coupled with humor (e.g., "what happens when water makes friends?")—a novel concept in a physiology textbook. Moreover, the text encompasses less than 200 pages, rendering it an easy read for students, residents, and fellows of any discipline and workload. I can enthusiastically recommend this text as an essential part of any medical curriculum that prepares clinicians of any discipline to care for the acutely ill or injured.

Lewis J. Kaplan, MD, FACS, FCCM, FCCP

The whole idea for this series arose from the physiology classroom and hospital teaching rounds. We realized that both in the classroom and on the wards, students and residents had a fair amount of knowledge regarding individual organ systems. However, there was still room for improvement regarding how all the organ systems integrate in order to respond to a particular situation. This book series is an attempt to bridge the gap of knowledge that divides organ from body, and isolated action from integrated response. It is our goal to have a series of books where integration of concepts serves as the primary focus. The books in the series are written so that they are hopefully easy to read, and that they can be read from beginning to end. It is our belief that if you truly understand something, you should be able to explain in a simple way. Therefore, we aim to tackle complicated topics with simple examples. And we hope that by the end of any book in this series, further more complex reading (e.g., the latest journal articles) should prove far easier to understand. We hope you enjoy reading these books as much as we enjoyed writing them.

Juan Pablo Arroyo

Adam Schweickert

DEDICATION

To our wives, Denise and Valentina, for their unwavering support of our every endeavor, both aimless and not so aimless.

ACKNOWLEDGMENTS

We wish to thank Mara Conner and Megan Wickline as well as the rest of the Elsevier staff. The amount of time and work that was dedicated to the making of this book is something we will always be grateful for.

We also wish to thank all those who provided their insight and suggestions throughout the writing of this book.

CHAPTER 1

How Fluid Is Distributed in the Body: Cells, Water, Salt, and Solutions

1.1 WHY IS WATER AMAZING?

We often tend to not give much thought to water. We drink it, we know we need it, we know the bulk of the earth is covered by it, but that's about it. When we talk about water in the body and its importance in chemistry, we think of it as sort of a boring liquid that just dissolves stuff. We seldom think about the key properties that make water so essential.

1. **It's everywhere**—Water covers about 70% of the earth's surface, it makes up 60% of the typical human body.
2. **It's polar**—Water, as we know, is made up of two hydrogen atoms and an oxygen atom covalently bonded. However, oxygen is more strongly electronegative than hydrogen. So, the mutual partnership of electrons is uneven: more of the electrons are on oxygen's side. This leads to a more negative charge on the oxygen side of the molecule and a more positive charge on the hydrogen side. This makes water a wonderful molecule for positive–negative interactions which help dissolve compounds (think electrolytes!).
3. **It's stable**—Water is stable with respect to heat change. Hydrogen bonding is largely responsible for water's very large heat capacity. Putting heat into a system where all the molecules are attracted to each other means it takes more energy to pull them apart. Conversely, it takes the removal of more energy for water to cool. This explains many naturally occurring phenomena! For example, it explains why coastal climates are more temperate, why sweating is an effective means of cooling (it absorbs more heat before evaporating from your skin), and why you only need to consume a relatively small amount of calories to maintain your body temperature.
4. **It's reactive (it plays well with others)**—Water knows how to both give and take. It is known as the universal solvent. This is due mainly to its size and polarity as discussed above. This means that

it tends to readily break the electrostatic attraction (positive–negative, magnet-type interactions) between stingy solutes. So, for example, when you put NaCl in water, it *ionizes* as it becomes Na^+ and Cl^-. So water's chemistry is necessary for many of the body's proteins to even function!

5. **It contains hydrogen!**—As mentioned earlier, water is small, polar, and highly concentrated within the body. Because ionization or hydration is involved in almost *every* biochemical process known, water is able to provide a nearly limitless supply of much-needed hydrogen ions!

1.2 WHAT HAPPENS WHEN WATER MAKES FRIENDS? BASIC PROPERTIES OF SOLUTIONS

A solution can be simply defined as liquid with other molecules mixed in. The mixture should be stable and uniform. The liquid, for example, water, is known as the **solvent**. The molecules mixed in, for example, table salt, are the **solute**. In general, the solvent is the thing there is more of in solution, and the molecules that are mixed in take on the properties of the solvent (e.g., a tablespoon of solid salt added to liquid water become liquid saltwater and looks just like pure water). When we talk about solutions in this book, we'll be talking about liquid mixtures, but understand that you can have gas and solid solutions as well, for example, air. Many of the properties of solutions are beyond the scope of this text, but we will detail what properties are clinically relevant and attempt to integrate them throughout the book.

1.3 QUANTIFYING SOLUTIONS

It should be noted that there are several properties of solutions that are entirely dependent on the *amount* of solute rather than what kind of solute. We can agree that a spoonful of salt added to a lake of water will be a different solution than a spoonful of salt added to a shot glass of water. But how do we quantify this difference? There are several ways, but the most useful way to quantify amount would be to know each solution's *concentration*. Well, chemists use a few different units to describe concentration. Two of the most pervasive are **molality** and **molarity**. The root of the word comes from an arbitrary quantity known as a **mole**. So, you should remember from basic chemistry that chemists love reactions. They also love mixing stuff. Let's say we want to play chemist and test this by making two different solutions, one with solute X and one with solute Y. We want to make the concentrations of each the same. The problem is that an individual X molecule weighs more than a Y molecule. But we can't count the individual molecules, and we can't just take 12 g of X powder and 12 g of Y powder and dump them into two equal glasses of water. If X molecule weighs twice as much as Y, then we'll have half the concentration of X molecules compared to Y! This led to the development of the mole. Thus, 1 mole of X = 1 mole of Y regardless of their individual weights.

Molarity = moles solute/L **solution**
Molality = moles solute/kg **solvent**

(Note that the arbitrary number was based on the number of atoms within a pure 12 g of normal carbon. This estimates to $\sim 6.022 \times 10^{-23}$ atoms/12 g ^{12}-Carbon = 1 mol).

1.4 FORCES AFFECTING STATIC SOLUTIONS

Ok, so that's solutions in a small nutshell. Now we're going to look at three very important concepts involving behaviors of solutions: **Diffusion, Gradients, and Osmosis**.

Diffusion means "to scatter, pour out." You should recall from your basic science classes that molecules—and thus matter—are made up of thermal energy. This energy is partly made up of self-propelled motion (i.e., kinetic energy). Well, back in 1827, a guy named Robert Brown, who was actually a Scottish botanist, was putting some pollen in a glass of water and watching it move around for hours (imagine his dinner party conversations!). He observed that this movement of pollen in water was random (later known as Brownian motion). People like Albert Einstein postulated that this applied to even smaller particles, including atoms. Turns out they were right. Particles will move constantly in random directions. Now imagine high concentration of these particles in solution. They will naturally collide and bounce off each other, right? You can imagine these particles will end up farther away from each other the more they collide, until they are dispersed evenly throughout the solution.

Now, let's imagine what would happen if you were to put a high-concentration solution in contact with a low-concentration solution.

As the high-concentration solution comes in contact with the low concentration, you would initially have a difference in the concentrations between solutions, this is known as a **gradient**. **A gradient is basically a difference (concentration, pressure, temperature, etc.) between two points.** In this example, the difference between the two solutions would be considered a **concentration** or **diffusion gradient**.

> **Key**
>
> Flow down a gradient is the movement from high concentration to low concentration!

This is easier to conceptualize and more applicable to the human body when you consider a large container separated by a semipermeable membrane (Figure 1.1A). Imagine the high-concentration solution on the left (A) and the low-concentration solution on the right (B). Which way do you think the particles (X) randomly wander toward? That's right! They'd randomly drift around and collide, and over a short amount of time, the particle collisions would drive them "down"

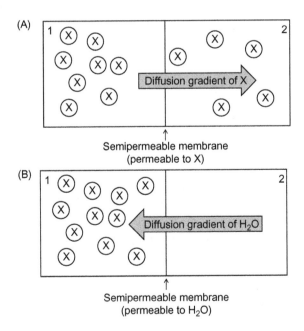

Figure 1.1 (A) Diffusion of X from compartment 1 to compartment 2 follows the gradient that exists between A (high concentration) and B (low concentration). (B) In this case, the semipermeable membrane is NOT permeable to X, but it is permeable to H_2O, thus water will diffuse from compartment 2 to compartment 1. In the end, in both A and B the ratio of $X:H_2O$ (i.e., concentration) will be the same.

the gradient to the area of lower concentration (from $1 \rightarrow 2$) (Note that this would only apply to the particles capable of traversing the semi-permeable membrane. Keep this concept in mind as we go along.).

Now knowing the above, **osmosis** becomes intuitive. Imagine a solution the same as before. Now make the membrane permeable to water and *not* to the other particles in solution (Figure 1.1B). (This is similar to what happens in the body, as we will see later on.) Water is made up of H_2O molecules that randomly wander just like any other particles despite their hydrogen bonding. If you imagine the water particles trying to move away from each other, they will move "down" the **osmotic gradient** from compartment 2 to compartment 1 (Figure 1.1B). Thus, the term osmosis is just a fancy way of saying: diffusion of water through a membrane.

To say it differently, water and solutes will try to achieve a balance! Water will try to distribute itself evenly throughout all compartments and solutes will try to do the same thing.

Note that the **osmotic pressure** is another way of stating this. If you look at Figure 1.1B again, the more concentrated the trapped solutes are on the left, the greater the "pull" that will be exerted on water to diffuse across the membrane. The "pull" is the osmotic pressure: the greater the "pull," the greater the osmotic pressure. This means that it is based on the **number** of molecules and *not* what kind of molecules they are.

Key

The osmotic pressure is determined by the number of molecules, not the size or charge. This is why 1 mol of NaCl has twice the osmotic power of 1 mol of glucose. (NaCl dissolves when placed in water, thus leading to 2 ions which will attract water.)

How do we express osmotic pressure in units? Because we established that osmoles are active, that is, they pull water, we divide the total number of osmoles by either kilogram of solvent (solid) or liter of solution (L).

Osmolality = osmoles/kg solvent
Osmolarity = osmoles/L solution

The terms osmolality and osmolarity are not the same. Osmolality refers to the total active osmoles per kilogram and would therefore be the proper term for use with regard to the body. However, you will find that this distinction is not always made, so keep an eye out for the difference.

1.5 DISTRIBUTION OF WATER

Ok, so we've looked at water alone. Now let's look at water in the body. To begin with, we need to understand how the water is distributed between body compartments. All organisms, including humans, are made up of cells and the principal component of cells is water! Water contributes to approximately 60% of total body weight in men and 50–55% in women (women have more fat content than men!). Total body water (TBW) is divided into intracellular (IC) and extracellular (EC) compartments. IC volume accounts for approximately 2/3 of the TBW and the EC accounts for the remaining 1/3. EC water is subdivided into two compartments, the interstitial space (IT) (i.e., the space between cells), which holds 2/3 of the TBW, and the intravascular compartment (IV), which contains 1/3 of the EC water (Figure 1.2). (Throughout the book we will be dealing with rough estimates and approximations, no need to memorize specific quantities, what's important is that you understand the concepts.) Note here that the majority of the water in your body is within cells, and what water is outside of cells is predominantly outside of the blood vessels. The distribution of water is not haphazard. What favors the 2/3 IC and 1/3 EC division of water is the concentration of solutes on both sides of

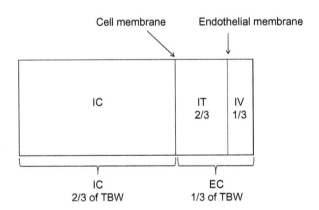

Figure 1.2 Distribution of water within the body compartments. Two-thirds of the TBW are located in the IC compartment, the remaining one-third is in the EC compartment which is further divided into interstitial (IT) being 2/3 of the EC and the IV compartment the remaining 1/3 of the EC water.

the cell membrane, and particularly the ion Na^+. Sodium is the most abundant EC ion, and it will play a key role in determining how much water is in the EC and how water distributes itself between IC and EC.

Clinical Correlate

In the hospital, one of the first treatment patients are often given are IV fluids, that is, fluids administered directly into the peripheral veins by a small, sterile plastic tube known as an intravenous (IV) catheter. That's all well and good, but as you recall, we're distributing these fluids within the smallest fluid compartment in the body! What good is done by inserting an IV if the fluid you're administering doesn't find its way *inside* the cells!? (Remember, it's *inside* the cells where the magic of life happens! The entire network of blood vessels in the body simply acts as a highway in service to the almighty and ubiquitous cell.) So, we need to get the fluids inside the cell ... but how? Thankfully, as you're about to read, nature provides an easy answer to this question.

Now, osmotic gradients are caused by different concentrations of fluid, right? But what exactly is concentration? How do we express it?

In a closed model like Figure 1.1, understanding diffusion is pretty straightforward, but in the context of the human body, we have to understand what constitutes the semipermeable membrane. The two major compartments of the human body are the EC and IC, but what divides the IC from the EC? The cell membrane! It is the cell membrane that will determine what is the final concentration of solutes on either side of its membrane. Interestingly enough, ions (e.g., Na^+, Cl^-, K^+, HCO^{3-}, and Ca^{2+}) and proteins CANNOT readily diffuse across the cell membrane. This leads to the accumulation of ions and/or proteins on different sides of the membrane. However, water IS permeable to the cell membrane, and it will follow its osmotic gradient. Therefore, water will follow the solutes, which in turn will be regulated by both the cells (particularly the Na^+/K^+ ATPase pump) and outside factors such as water and solute intake. Now, if we look back at Figure 1.2, the cell membrane is not the only membrane separating water in the body. The **endothelial membrane**, the membrane lining the outside of the cells that make up the blood vessels is also a key factor as it separates the two ECF components, namely blood plasma and interstitial fluid. The endothelial membrane, unlike the typical cell membrane, is freely permeable to not only water but also to glucose and ions. However, it is NOT readily permeable to proteins or larger things like cells! Thus, cells and protein will remain "trapped" in the plasma volume. This makes sense, as the blood vessel's role is to deliver nutrients and remove waste, right? So, it would help if glucose, electrolytes, and lipids could be easily transported to and from the cells that are on either side of the endothelial membrane.

To recap we have:

1. Cell Membrane
   a. Divides IC from EC
   b. Permeable to water
   c. Not readily permeable to electrolytes, glucose, or proteins
2. Endothelial Cell Membrane
   a. Divides IV from EV
   b. Permeable to water, electrolytes, and glucose
   c. Not readily permeable to proteins

An example of endothelial function in the clinical setting relates to inflammation. One of the primary signs is *tumor*, or swelling. This occurs because the local endothelial membrane becomes more permeable to proteins and cells (which are important to the "war effort"). This causes water to follow the gradient created by out of the vascular space and into the extravascular space! (Protein permeability is increased during inflammation, however, for the remainder of the book, we will consider the endothelium as impermeable to proteins.)

1.6 FORCES AFFECTING STATIC FLUID IN THE BODY

So, the cell membrane functions as our semipermeable membrane through which water, but no solutes, can diffuse.

Key

Water can diffuse across the cell membrane but ions and proteins can't.

As we mentioned earlier, water and solutes will try to find equilibrium, in this case between the EC and the IC compartments. So, water movement in the body is going to be driven by the differences in solute concentration!

But this whole discussion is pointless if we don't have a semipermeable membrane that can allow for solute concentration on either side. How do we define the force that drives the movement of water when you're properly "holding something back" (i.e., solutes)?

Let's begin with osmolality.

Osmolality is the TOTAL concentration of solutes, both permeable and nonpermeable in a solution. It is expressed as osmoles per liter. (An osmole is a molecule that can pull water.) Osmolality is a property of the solution by itself! That is, the osmolality of an apple juice is approximately 800 mOsm/L, while the osmolality of pure distilled water is 0. The normal osmolality of the human body oscillates between 280 and 300 mOsm/L.

Another term that is frequently confused with osmolality is tonicity. **They are not the same thing!**

Tonicity is the concentration of *effective* solutes. It is "effective osmolality." It is another way of expressing solute concentration much like osmolality, but it only includes those solutes that CANNOT cross a *specific cell membrane* (i.e., solutes that exert an osmotic force). Look at Figure 1.3. Because tonicity is basically "effective osmolality," it is a more clinically-relevant term. When we talk about tonicity in medicine, the specific membrane is the red blood cell (RBC) membrane. If the plasma outside the RBC membrane has more impermeable solutes than the fluid within the cell, then the plasma or EC fluid is said to be hypertonic with respect to the IC fluid (EC tonicity > IC tonicity). If it is lower, it is called hypotonic (EC tonicity < IC tonicity). The take-home message is that tonicity is a relative term because it will change depending on the membrane that is being used as a reference. For all intents and purposes, our reference membrane will be the RBC membrane.

In order to understand how water and solutes move within the body, let's take a look at our starting point. Figure 1.4 is a schematic representation of the distribution of ions and proteins in the different spaces. As you can see, Na^+ is the most abundant EC ion. As it is the most abundant EC ion and it has osmotic activity, this means that it pulls water! Thus, it is the total **QUANTITY** of Na^+ that determines EC volume. While the **CONCENTRATION** of Na^+ (the ratio of Na/H_2O in the EC space) will determine the fluid shifts between IC and EC. Let's analyze fluid shifts a bit further.

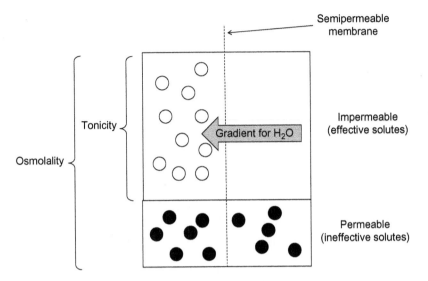

Figure 1.3 Tonicity vs. osmolarity. Tonicity is "effective osmolality" in that if a molecule is permeable to a specific membrane (black dots) it does not add to the tonicity, but it does add to the total osmolarity. However, if a molecule is impermeable to a specific membrane it contributes to both tonicity and osmolarity.

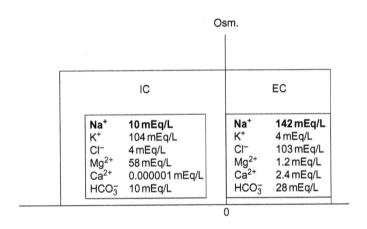

Figure 1.4 Differences in ion concentrations between IC and EC.

Key

The major EC ion is Na^+. Because Na^+ is NOT readily permeable to the cell membrane, it will be the major determinant of EC volume and consequently IC volume. The total Na^+ content (i.e., how much Na^+ is there) will determine EC volume, while Na^+ concentration (i.e., how much Na^+ in how much water) will determine cell volume!

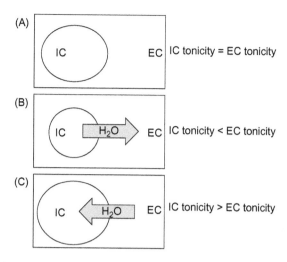

Figure 1.5 Net diffusion of water is dependent on the differences in IC and EC tonicity. If we consider the IC osmolarity as our reference value and then compare the tonicity between IC and EC, in equilibrium (A) there will be no net water movement. If EC tonicity increases (B) water will shift from the IC to the EC. If EC tonicity decreases (C) water will shift from the EC to the IC

Figure 1.5 shows us what happens to the flux of water with changes in the EC tonicity. In equilibrium, the concentration of solutes in both the EC and the IC are the same, and thus the EC is considered isotonic (iso = the same, see Figure 1.5A). With equal tonicities, there will be no net solute gradient and no movement of water. However, an increase in EC tonicity (Figure 1.5B) (making the EC hypertonic compared to the IC), will favor diffusion of water from the IC to the EC following the solute (tonicity) gradient. Conversely, a decrease in EC tonicity (Figure 1.5C) will drive water into the cell. You'll notice in the figure that there are volume changes in the cell secondary to the movement of water. In the human body, cell volume has to be maintained relatively constant. Any significant change in cell volume will significantly alter cell physiology and hence the normal body functioning. So, the body has developed ways to counteract both acute and chronic changes in EC tonicity to prevent changes in cell volume. In short, if a cell is challenged with increased EC tonicity, it will draw in electrolytes to offset this unbalance and reclaim the water it initially lost, this is called regulatory volume increase. On the contrary, if a cell is challenged with a decreased EC tonicity, it will shuttle electrolytes out in order to shed the excess water, this is called regulatory volume decrease.

Clinical Correlate

Mannitol is a type of drug that is used to treat cerebral edema. The cell wall is impermeable to mannitol, therefore mannitol draws water out of the cell by increasing the tonicity of the EC fluid; this will drive water out of neurons, shrinking them, and ameliorating the edema.

Taking the new concepts of tonicity and osmolality into account, we can create a new model for representing the distribution of water in the body. Figure 1.6 is a graphical representation of the fluid distribution in the body with regard to total body osmolality. It is called a Darrow–Yannet diagram (we'll abbreviate this as D–Y diagram), after the authors who first proposed the figure in 1935. IC volume increases/decreases are represented by changing the length of the boxes representing IC or EC on the left and right, respectively. Changes in osmolality are represented by changes in the height of the boxes. We can use the D–Y diagram to see what the changes in one compartment would do to the changes in the other compartment. (The initial state will be represented by solid lines, dotted lines will represent changes in either osmolality or the water volume of each compartment.) For the time being, we will consider the EC to be a unified space (we will discuss the difference between IV and interstitial fluid later on). We will start with the EC when looking at changes because when you gain or lose fluid it starts with the EC.

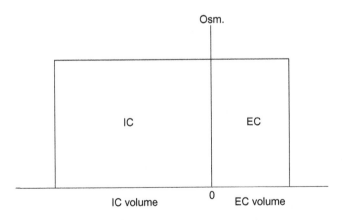

Figure 1.6 Darrow–Yannet diagram. Relation between IC and EC volume with osmolarity. The x-axis represents total volume of either IC or EC. The y-axis represents osmolarity.

Think of an IV catheter. What would happen if we expand EC volume without changing the tonicity of the fluid? This scenario is something relatively common in hospitals. The infusion of an isotonic solution, that is, solutions that have the same amount of osmotically active solutes as the EC fluid, is one such example.

Clinical Correlate

The two basic need-to-know isotonic solutions are normal saline or (0.9% NaCl solution), which has 154 mEq of NaCl per liter of H_2O, and Lactated Ringer's solution which has 130 mEq of Na^+, 109 mEq of Cl^-, 4 mEq of K^+, 28 mEq of Lactate, and 3 mEq of Ca^{2+}. (Remember that these solutions are isotonic when compared to plasma! If we were to compared them to apple juice they would be hypotonic.)

In the case of an EC fluid expansion that does not change the osmolality of our compartments, there would be NO fluid shift, as illustrated in Figure 1.7 where we can see an increase in EC fluid, that does not change the osmolality of the entire system or the IC volume. Therefore, an isotonic increase in EC volume is distributed only in the interstitial fluid and the IV volume. Likewise, an isotonic decrease in EC volume (Figure 1.8) will decrease the EC volume, thereby decreasing interstitial and IV volume without affecting IC volume. Remember that solutes are not freely permeable across the cell membrane, but water **is** freely permeable. The take-home message from all of this is the following: **in order to move fluid in and out of the cell there needs to be a change in EC fluid osmolality which will generate a gradient for the movement of water.**

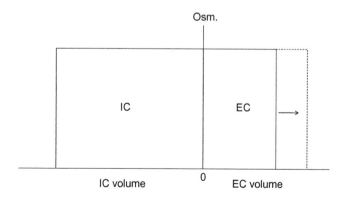

Figure 1.7 Isotonic increase in EC volume represented by the dotted line. Note how this increase does not affect IC volume or the osmolarity.

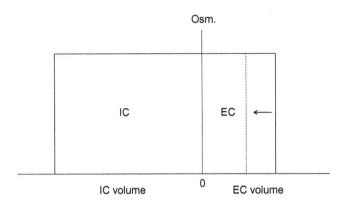

Figure 1.8 Isotonic decrease in EC volume represented by the dotted line. Note how this decrease does not affect IC volume or the osmolarity.

Key

The term EC osmolality is used to discuss changes in EC vs IC volume because the effective osmolality and the total osmolality (tonicity) of the EC fluid are almost equal.

Fluid shifts due to changes in EC fluid osmolality can be explained using the D−Y diagram. As we can see in Figure 1.9, an increase in EC osmolality will shift fluid from the IC compartment to the EC compartment. So, an increase in EC osmolality will deplete cells of their volume (triggering regulatory volume increases, see above). This fluid shift will occur in three steps (labeled arrows in Figure 1.9). First there will be an increase in EC osmolality (1), fluid will then shift out of the cells following the "flow down gradient" for water (2), and into the EC thereby expanding EC volume at the expense of IC volume (3), which ultimately leads to an increase in IC osmolality achieving balance once again. This is an interesting scenario because of the compensation that needs to occur in order to offset the initial increase in EC osmolality, that is, to diminish EC osmolality by either losing solutes or gaining water. Remember that the body will always try to maintain "homeostasis" which means that a system will try to maintain a stable internal environment in the presence of a shifting external environment. This means the body will attempt to compensate whatever happens outside. (We already saw an example of this in regulatory volume decreases/increases in which cells will try to regulate their volume

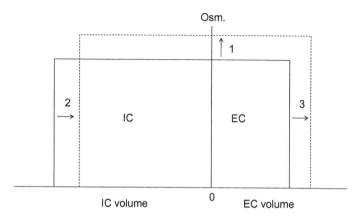

Figure 1.9 An increase in EC osmolarity (1) will shift water out of the cell (2) and into the EC volume (3), allowing the IC osmolality to be in equilibrium with the EC osmolality (4).

in the presence of changing EC tonicity.) So, if the body will try to minimize change, what do you think is going to happen as a result of the increase in EC tonicity? The body has one of two options: (1) retain solute-free water, thereby diluting the EC compartment and shifting water back into the cells or (2) excrete solutes, which will decrease EC osmolality and shift fluid back into the cells.

What happens if there is a decrease in EC osmolality? In Figure 1.10, we can see the effects of a decrease in EC osmolality in the D−Y diagram. Once again it is divided into phases. Initially there will be a decrease in EC fluid osmolality (1). This decrease in EC osmolality will drive water out of the EC fluid and into cells (2). The water will then enter cells increasing IC volume and decreasing IC osmolality (3). What compensations have to occur here in order to balance out the initial problem? Well, let us analyze this a little differently, what is wrong with our diagram? The osmolality is decreased, the IC volume is increased, and the EC volume is decreased. What could we do to solve this? (Remember that in the clinical setting you only have access to the EC compartment through an IV catheter and that the IC responds to whatever is happening in the EC, so we need to do something to the EC.) Could we infuse water into the EC compartment? This would not help us because it would further decrease EC osmolality (we would actually be making the problem worse). How about we increase the EC solute concentration? This could work! Why? Because this is correcting our initial problem, if we infuse solutes

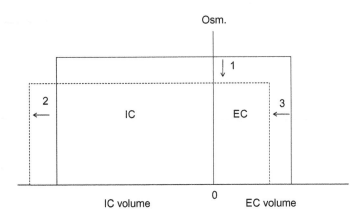

Figure 1.10 A decrease in EC osmolarity (1) will drive fluid into cells (2) and contract EC volume (3).

into the EC compartment we will increase EC osmolality, thereby shifting fluid from the cells into the EC compartment. What if we eliminate water from the EC compartment? At first sight this might seem like a good idea, because if you decrease the water content of the EC compartment you would increase osmolality, thereby shifting fluid from the cells back to the EC. However, if we analyze this closely this will be a bad idea, why? Because if the EC volume is already depleted (remember water got shifted from the EC to the IC) depleting it even further by depriving it of water would shrink plasma volume (we will get into this a little later, but for the time being remember that blood and plasma volume directly relate to EC volume), and if we deplete plasma volume our patient could be in serious trouble!

1.7 CLINICAL RELEVANCE

From our last example, we were able to see how alterations in EC/IC changes could impact a patient, now let's get into this discussion a little further. We stated at the beginning of the chapter that the EC compartment was divided in two: the interstitial compartment, which accounts for 2/3 of the total EC volume and the IV volume, which accounts for 1/3 of the total EC fluid (Figure 1.1). Therefore, any changes in the EC compartment will affect IV volume! Before we go into the exact fluid shifts, let's understand what blood is composed of Figure 1.11 shows us the components of blood. It can be divided into two: plasma volume and cellular components. Plasma volume accounts for approximately 55% of total blood volume and it is subdivided into two: water which is roughly 90% (this is the water that will be shifting

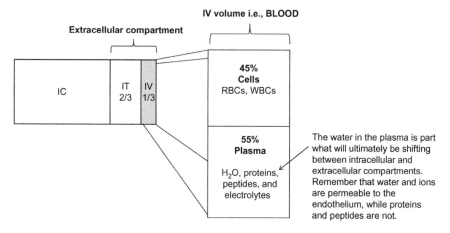

Figure 1.11 Blood is divided into plasma and cellular components. Plasma is equivalent to 55% of the blood volume and is roughly 90% water and 10% extras (proteins, nutrients, hormones, and electrolytes). Cellular components are 45% of the total blood volume, and it contains both RBCs and WBCs.

back and forth between the IC and the EC) and what we will call additional components: electrolytes, plasma proteins (albumin, fibrinogen), and hormones (peptide, amino acid-derived, and steroid hormones). Cell components are the remaining 45% of total blood volume, and it's mainly made up of RBCs and white blood cells (WBCs). (Remember though that all cells in the cellular compartment are a part of the IC volume! So whatever happens to the total IC volume is happening to the RBCs and WBCs and to the rest of the cells in the body later as the blood is pumped and the fluids diffuse throughout the capillaries in your body).

Aside from blood, the interstitial space deals with the remaining 2/3 of the EC compartment. This is the water that is located between cells, and it is also from where the lymphatic system drains its fluid. Fluid that is contained in the interstitial space is different from the fluid contained in the blood (Figure 1.12). Why? Because blood is inside blood vessels (arteries and veins) and both arteries and veins are made up of three layers of cells (the intima, muscularis, and adventitia layers) which divide it from the IT space. In the capillaries where the layers of the blood vessels are the thinnest, fluid can leak out, but it is a special type of fluid devoid of proteins! The only thing that will leak out is H_2O and electrolytes! (So you see, blood vessels, and more specifically capillaries, are a special kind of semipermeable membrane through which water AND electrolytes can leak through). This characteristic

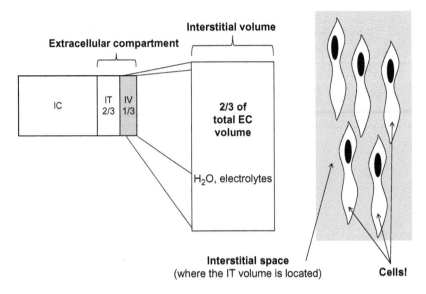

Figure 1.12 Interstitial fluid is devoid of proteins but is rich in H_2O and electrolytes.

provides us with the reasons behind our major divisions of IC and EC volume (see discussion regarding Figure 1.1).

Now that we know what fluid will be moving from the EC into the IC, that is, plasma volume and interstitial fluid, and that plasma volume is dependent on EC volume, let's try and put together the pieces of the puzzle we've been building in the clinical context. We will set out a couple of examples along with the corresponding D−Y diagram.

Key

In order to properly evaluate a board question regarding a D−Y diagram, you have to take three basic steps: (1) Is there a net fluid gain or a net fluid loss? (2) Is the EC osmolality changing? (Remember that if EC osmolality changes there is going to be a fluid shift.) (3) If EC osmolality did change, is water going to go in or out of the cells?

1.8 CLINICAL VIGNETTES

Example 1

A 20-year-old football player was found unconscious in the locker room. He had been exercising in full gear with little access to water. The outside temperature was around 32°C (89.6°F).

1. What fluid distribution problem is present in this patient?
   - In this clinical context, we can see that our patient has been doing heavy exercise in full gear (which means he probably has been sweating a lot due to heat). Sweat is hypotonic, that is, it has more water than electrolytes. Also, you can sweat a lot! A typical adult can lose a significant amount of their body weight in water loss due to sweating with prolonged, intense exercise! If he had restricted access to water, then our patient has been losing sweat, that is, hypotonic fluid! So, the problem is a: hypotonic fluid loss!
2. What would the D–Y diagram associated with this patient look like?

   In order to understand the D–Y diagram, we have to think in sequential steps:
   - *Is our patient losing fluid or gaining fluid?* Losing fluid, so there's going to be a decrease in total volume.
   - *Is the EC osmolality going to change?* It is going to increase. So, there is going to be a fluid shift because there's a change is osmolality. If EC osmolality increases what is going to happen to the IC volume? It is going to shift from the IC compartment to the EC compartment (which remember is already depleted). This is depicted in Figure 1.13

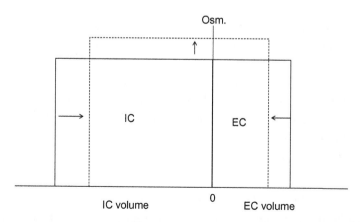

Figure 1.13 D–Y diagram of hypotonic fluid loss. Solid line represents steady state, the dotted line represents the outcome of losing hypotonic fluid, for example, profuse sweating. EC and IC total volume decreases and total osmolarity increases.

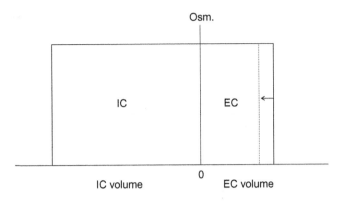

Figure 1.14 D−Y diagram of isotonic fluid loss. As there is no change in osmolarity, there is no fluid shift, therefore only EC volume is depleted. (Remember that blood is a part of the EC volume).

Example 2

A 32-year-old female is the unrestrained driver of a motor vehicle collision. She is found at the scene with an exposed fracture of the right femur and substantial blood loss.

1. What is the fluid distribution problem in this patient?
   - Our patient has just suffered massive trauma and is losing blood quickly. Blood however is isoosmolar with respect to IC fluid, that is, in a steady state the osmolality of blood and of cells is the same. So, this case represents: isotonic fluid loss!
2. What would the D−Y associated with this patient look like?
   We will follow the same steps as from the example above.
   - *Is our patient losing or gaining fluid?* Losing fluid from the EC volume.
   - *Is the EC fluid osmolality going to change?* NO!
   - *Is there going to be a fluid shift?* If the EC fluid osmolality is not changing, there will not be a fluid shift. This is depicted in Figure 1.14.

Example 3

A 45-year-old female undergoes an elective cholecystectomy (i.e., her gallbladder is taken out, in a nonemergent setting). During the surgery, there is significant blood loss (approximately 1 L), but she is infused 3 L of normal saline IV solution.

1. What is the fluid distribution problem in this patient?
   - This patient just had surgery in which she lost 1 L of blood. However, she was infused 3 L of normal saline (remember normal saline or 0.9% NaCl solution is an isotonic solution), this leaves us with a total gain of 2 L! She just had an isotonic fluid gain!
2. What would the D−Y associated with this patient look like?
   - *Is this patient losing fluid or gaining fluid?* Losing EC fluid.
   - *Is the EC fluid osmolality going to change?* No. Remember that normal saline is isotonic.
   - *Is there going to be a fluid shift?* No, there is no change in the total osmolality. This is depicted in Figure 1.15. However, a particular detail with this example is the following: blood is not the same as normal saline. Normal saline lacks proteins! From our discussion regarding interstitial fluid, we said that blood had proteins that the interstitial fluid lacked. Proteins have the capacity of attracting water and keeping it in the IV space, but we're replacing blood (that has proteins) with an isotonic solution that lacks proteins! From this we can infer that our normal saline solution will be distributed not only in the IV volume but also in the interstitial volume! (If there were proteins in our solution, it would stay in the IV compartment; because there are none, then our solution will be distributed in the entire EC volume compartment, and remember the EC is 1/3 IV and 2/3 interstitial.) So, our solution is going into the EC compartment,

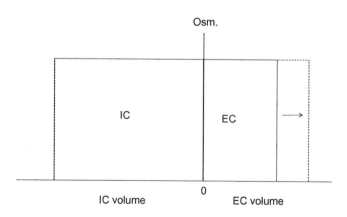

Figure 1.15 D−Y diagram of isotonic fluid gain. Increasing the EC volume without changing total osmolarity would result in EC fluid expansion without changing IC fluid.

how much is going into the IV and how much into the interstitial? Well, if the IV is 1/3 of the EC and the IT is 2/3 of the
EC, and our solution is being distributed evenly in the EC,
then 2/3 will go to the interstitial compartment and 1/3 will go
to the IV compartment, roughly 1 L divided in thirds yields
667 mL in the IT (2/3 of 1 L) and 333 mL in the IV (1/3 of L).
This means that whenever we give a patient isotonic IV fluids,
we will be expanding both the IT and the IV spaces!

Key

The administration of isotonic IV fluids increase both the IV space and
the interstitial space.

Example 4

A 60-year-old male patient confuses his daily aspirin tablets with
concentrated NaCl tablets; he notices his mistake after 10 days.

1. What is the fluid distribution problem in this patient?
   - This patient is a little different from our previous patients. In
     this case, there is no net fluid loss or gain. Instead, what we are
     modifying is the osmolality! Why? Well, our patient began
     ingesting ridiculous amounts of salt. If salt is not freely perme-
     able between the EC and the IC, where is the salt going to be
     distributed? In the EC fluid! So, our fluid distribution problem
     is actually an increased EC fluid osmolality!
2. What would the D−Y diagram associated with this patient look like?
   - *Is this patient gaining fluid or losing fluid?* None, there is no net
     gain or loss of fluid. This means that the total amount of fluid
     in the body will stay constant.
   - *Is the EC osmolality going to change?* Yes! If this patient is
     ingesting large amounts of salt, the EC osmolality will increase.
   - *Is there going to be a fluid shift?* Yes! If the EC osmolality is
     increased, then fluid will shift from the IC to the EC compart-
     ment. The D−Y diagram is depicted in Figure 1.16.

Example 5

A 25-year-old female patient starts drinking 10 L water low in
NaCl a day.

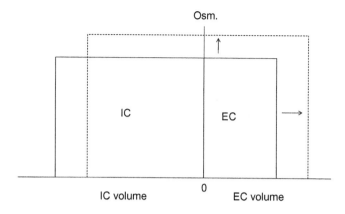

Figure 1.16 D–Y diagram of EC osmolarity increase. The increase in EC osmolarity will shift water from the IC to the EC without changing TBW content.

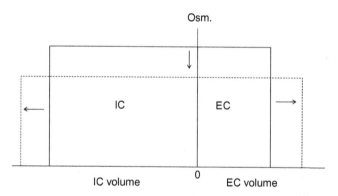

Figure 1.17 D–Y diagram of an increase in hypotonic fluid. By increasing hypotonic fluid, EC osmolarity will decrease. This will shift water into cells, thereby decreasing IC osmolarity and increasing IC volume.

1. What is the fluid distribution problem in this patient?
   - This patient is gaining large amounts of hypotonic fluid. She is probably diluting herself!
2. What would the D–Y diagram associated with this patient look like?
   - *Is this patient gaining fluid or losing fluid?* This patient is obviously gaining large amounts of hypotonic fluid!
   - *Is the EC osmolality going to change?* Yes! If there is a large increase in water, the EC osmolality is going to decrease.
   - *Is there going to be a fluid shift?* Yes! If EC osmolality decreases in the face of a large gain of hypotonic volume, then water will shift into the IC and out the EC. The D–Y diagram is depicted in Figure 1.17.

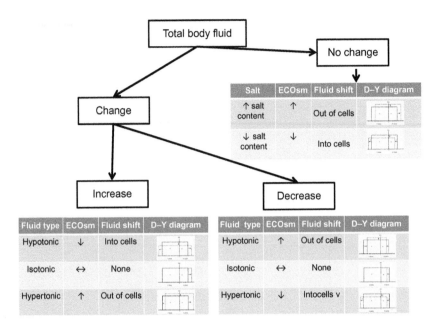

Figure 1.18 The three step approach to solving variations in total body fluid and osmolarity. (1) Is there any net change in total quantity of fluid? (2) Did the salt (Na⁺) content increase or decrease. (3) Is water going to move in and out of cells or not at all?

Figure 1.18 presents a quick summary of how to approach changes in total body fluid.

Throughout this chapter we have reviewed several key concepts including: diffusion, osmosis, tonicity, osmolality, how fluid is distributed through the body and how this fluid distribution and shifts is dictated by EC and IC osmolality. However, we have seen this from a static point of view, in the next chapter we will analyze fluid movement from a dynamic point of view.

CHAPTER 2

Fluid Movement in a Rigid Tube: Pressures and Gradients

2.1 FLUID DYNAMICS

In the last chapter, we focused on the distribution of fluid in the body and how certain forces cause water to move on its own without a pump. We were dealing with lab containers, artificial membrane dividers, and theoretical cells, but we could've just as easily been working in a bathtub divided by a towel! Now we're going to add another variable. In this chapter, we're going to look at concepts guiding fluid movement on a larger scale. We are still looking at movement without a pump (i.e., without the heart). But now, we're going to analyze some of the forces that come into play when the environment starts interacting with the water, as well as what happens when water starts to move. Bear in mind that all of the concepts from Chapter 1 are still happening behind the scenes, and at time these forces will be at odds with one another.

Now, remember that for osmosis to occur, it requires a semipermeable membrane. Well, in the body, the capillaries are semipermeable. The big fat arteries and veins are not! These act more like a pipe or a hose. So, keep in mind that larger vessels MOVE the blood and the capillaries exchange it. The movement of blood is called blood flow. It is through blood flow that oxygen is delivered to the tissues; thus, we need to understand the basics of flow in order to later understand how end-organ perfusion and diffusion across the capillaries take place.

As mentioned before, fluid flow is the continuous movement of a certain amount of fluid from one point to another. Flow is defined as volume of fluid moved over a time period. Now, we could get fluid to flow simply by letting gravity carry it, for example, pouring a pitcher of water upside down. But that wouldn't work so well if you were trying to get fluid to flow against the direction of gravity, such as in the body, where blood has to move up to the brain and back down, for example. Without some other way to drive blood flow, you'd probably have to do cartwheels continuously just to live! So, let's look at how we can generate and predict fluid flow.

One way to generate fluid flow is to generate a change in pressure. If I take my mouth and blow into one end of a hose filled with water, water will come out the other end. What is more, we can predict exactly how much. In this instance, two things determine the amount of fluid flow: a change in pressure and the resistance within the tubing. You can estimate how much water will come out the other end by taking the equation Flow $= \Delta P/R$, where ΔP is the change in pressure and R is resistance (note that this is a variant of Ohm's law you may have learned in introductory physics when talking about flow of electricity). So let's start with the driving force behind flow: pressure. Remember our concentration gradients from Chapter 1? Well, if we generate pressure at one end of the tube and there is less pressure at the other end (because no one is blowing air into that side), we've generated a pressure gradient. There is more pressure on one side than the other. But what exactly is pressure?

2.2 CONCEPT OF PRESSURE

If you were to cut something with a knife, would you use the sharp end or the flat end? Pressure is considered the force per unit area. The mechanics of pressure with a solid is simple. If we push down with the knife, the force is applied in a direction perpendicular to the surface of the object we're cutting. The sharp edge has less area to it than the sides of the knife, so we generate more pressure with the same amount of force. Therefore, if we wanted to generate more pressure, using the same amount of force, we'd use the sharp end. Solids are a little more straightforward because the molecules are tightly bonded together. If we push down perpendicular to the object we're cutting, we create a pressure force in that same direction. However, with a liquid or a gas, pressure becomes a little bit trickier. If we have a container filled with a gas, the molecules in the middle are banging around and experiencing forces in all different directions, but the molecules up against the wall are banging into the walls perpendicular to the inside wall of the container. If we looked at a very small part of this container wall over a short amount of time, we would see that the molecules hit at random angles. But if we were to watch long enough, we would see that, on average, the force is perpendicular to the surface. This is true of all the surfaces of the container. The force acting on the wall of the container is what we call pressure. Now, if we were to try and compress a gas by shrinking the container or stuffing more gas in, we could do so. In the meantime, we would increase the pressure inside the container in all of these perpendicular directions as more frequent and more forceful collisions would happen between the gas molecules inside the container. Fluids, however, are not compressible! If you try and compress a fluid in a container, the liquid molecules will hit up against the container wall in a perpendicular fashion just like with gases, but because the liquid molecules are more tightly bound together, there's no extra room for more. So when you attempt to compress a liquid in a container you get even larger amounts of pressure (Figure 2.1)! The pressure felt against the container by this fluid, particularly when at rest, is called hydrostatic pressure (*hydro* = water; *static* = at rest). Any fluid that's in a container would leave that container if the container had a hole in it. The pressure that exists as a result of the fluid **NOT** being able to escape is called hydrostatic pressure.

Ok, great, but why are you wasting time reading about this? Well, now you understand blood pressure! Blood pressure is the pressure

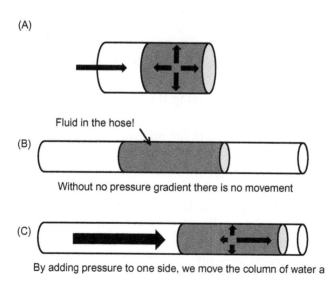

Figure 2.1 (A) When we exert pressure onto a fluid in a container, the force is exerted in all directions. (B) The pressure is evenly distributed in all directions. In a nondistensible container, the addition of force (large arrow). (C) will drive the movement of fluid from left to right.

exerted by the blood against the walls of its container: the blood vessel! Blood pressure is hydrostatic pressure!

2.3 BLOOD PRESSURE

In medicine, blood pressure is the pressure exerted by blood against the walls of the blood vessels, especially the arteries. One of the most important discoveries in medicine was that you can measure blood pressure by compressing the arteries by wrapping a cuff around the arm and squeezing until you no longer have detectable blood flow through the artery. Through slowly releasing the pressure around the arm by deflating the cuff, you can measure the point at which blood flow returns. Much like in Figure 2.1, where you have a force exerting pressure on liquid in a glass cylinder in all directions, the same applies to blood in the major arteries with each heartbeat. Some of this pressure pushes the blood further down the pipe, but some of this pressure pushes outward against the vessel walls to help keep the arteries distended. This pressure against the wall is stronger when the heart contracts (called systole) and weakest when the heart relaxes (called diastole). But even in diastole, there is enough blood exerting pressure against the arteries that it still acts to help keep the arteries open. When they first started using these cuffs to measure blood pressure in the late 1800s, they used their fingers to palpate for loss of and return of pulse, giving a single number for maximum blood pressure. It took about 10 years, but a Russian vascular surgeon by the name of Nikolai Korotkoff realized you get key pieces of information if you use your stethoscope! By squeezing the cuff tight and slowly releasing the pressure, you can measure the point at which you hear blood rush forward. This is the point where the internal distending pressure generated by each heartbeat is strong enough to overcome the external compression pressure of the cuff, which corresponds to the systolic blood pressure. Continuing to release the cuff pressure, you come to point where you hear no sound at all with your stethoscope. This is the point where the pressure in the cuff isn't even enough to compress the artery when there is no force being generated by a heartbeat and corresponds to the diastolic blood pressure. Try "compressing" the hole formed when you open your mouth. Now blow. Slowly open your mouth wider and continue to blow. The noise you're hearing is caused by obstruction of airflow. The same is true for obstruction of blood flow, only it can't hold a melody!

2.4 PRESSURE AND FLOW

Now, the blood vessel container can distend and contract somewhat, much like a water balloon that you can squeeze, this capacity to expand and contract has a huge effect on flow, so to make things simpler for our initial discussions about fluid mechanics, we're going to assume a rigid container: a glass cylinder (Figure 2.2). Water will be able to flow through our glass cylinder depending on the pressures inside the cylinder. If pressures are different at opposite ends (either more fluid is added or we attempt to compress the fluid at that end), then we have create a pressure gradient. This is like a concentration gradient in that it too has a potential energy associated with it. If we fill up one end with lots of water, those molecules are banging around against all the sides on that end. If there is less pressure on the other end, then the molecules will quickly migrate where there are fewer collisions. Essentially, water pushes off of water. The pressure difference or pressure gradient is denoted ΔP. This can be viewed mathematically as: $\Delta P =$ pressure at point A $-$ pressure at point B or $\Delta P = A - B$. Changing the pressure values will have any of three outcomes:

1. A > B—This means that the pressure at point A is higher than the pressure at point B. If this is the case, then water will flow down the gradient from an area of high pressure to an area of low pressure, that is, from point A to point B.

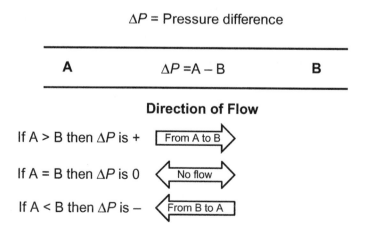

ΔP = Pressure difference

Figure 2.2 ΔP represents the pressure difference between two points (A and B). The pressure gradient will establish the direction in which flow will occur. Water will flow from an area of high pressure to an area of low pressure.

2. A = B—This means that the pressure at point A and B is the same. If there is no pressure difference, then there is no pressure gradient and no flow will occur.
3. A < B—This means that the pressure at point B is higher than the pressure at point A. If this is the case, then water will flow down the gradient from an area of high pressure to an area of low pressure, that is, from point B to point A.

The take-home message is that the direction of the movement is determined by the pressure gradient, and it is from an area of high pressure to an area of low pressure.

Key

The movement of blood will be from an area of high pressure to an area of low pressure.

2.5 RESISTANCE

Resistance is the second determinant of flow. Resistance (R) is defined as the opposition to something. Just think if you're resisting something, you're opposing it. In the case of blood flow, vascular resistance does just that, it opposes blood flow, thereby reducing the total **quantity** of blood that is going through a blood vessel at any given moment. Figure 2.3 shows us what resistance in our "glass cylinder" model looks like. It is basically a narrowing of the diameter of our cylinder. Think of resistance in the following way: what happens to the traffic on a four-lane highway when there's construction and two of the four lanes are blocked? Traffic slows down (i.e., the traffic FLOW slows down!) because half the number of lanes can go through. Now, all the four lanes have to merge into two in order to get past the area of resistance.

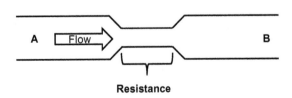

Resistance

Figure 2.3 Resistance is the opposition to flow. In this case, the decrease in diameter between point A and B opposes the flow.

So, we stated that the two determinants of flow were change in pressure and resistance. How does this work exactly? The more pressure there is in point A vs point B, the larger our ΔP (ΔP = pressure A − pressure B). The larger the ΔP, the more flow there is going to be. These are two concepts that are directly related, as the ΔP rises, flow rises as well! Figure 2.4A shows us this linear relationship. However, resistance is a little different. We have stated that resistance opposes flow, so as resistance increases flow will decrease. Resistance and flow, however, are inversely related (again Flow = $\Delta P/R$.) As R rises, flow decreases as we can see in Figure 2.4B. This relationship is **not** a linear one. Changes in resistance, as you can see in the figure, have much bigger impact on the overall flow than changes in pressure. With our traffic analogy, this inverse relationship makes sense intuitively.

Imagine for a moment ΔP as the change in the number of cars driving on our four-lane highway from 5:30 pm on a Sunday compared to 5:30 pm on a Monday evening. Even though everyone may be driving a little slower with all the added cars as people are making their way home from work, assuming no major delays, if you were to stand on the side of the road and count the cars that pass by over a minute's time you'd see that the traffic flow was much greater on Monday. Much like adding more fluid to a tube, or blood to a blood vessel, adding more cars all heading in the same direction causes an increase in ΔP, and thus an increase in the flow.

Now look at that same example, but assume there is a construction closing down two lanes. So now what would you expect? That's right, a traffic jam. It doesn't matter if it's Sunday or Monday,

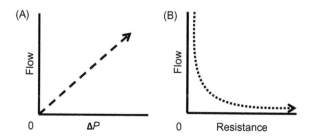

Figure 2.4 (A) Linear relation between pressure and flow. As pressure increases, flow increases. (B) Exponential relationship between resistance and flow. As resistance, increases flow decreases.

there's still likely going to be a bit of a traffic jam. As people start tapping their brakes trying to merge, you will see a ripple effect as cars slow down more and more. Now, remember our curve from Figure 2.4B. Even small changes in resistance cause a huge decrease in flow. All we did was take away two lanes. We can increase our pressure gradient all we want, but there is only so much flow through that construction zone.

Key

As the ΔP increases, flow increases. As R increases, flow decreases. And, flow is affected more strongly by a change in resistance than by an equal (but opposite) change in pressure.

Ok, so why are we spending all this time talking about cars? Because the exact same principles hold true for blood flow! In the 1800s, a French physician named Louis Poiseuille observed this traffic phenomenon in blood flow. This guy had it all: looks, smarts, and a fancy name. In 1840, he formulated and published Poiseuille's law. This law states that $R = 8nl/\pi r^4$, where R = resistance, n = viscosity of fluid, l = length of the blood vessel, and r = the radius. What's cool about this equation is that if you look carefully, it basically gives you the recipe for resistance. You'll notice that tube **radius** is by far the **largest contributor to resistance**! Even a seemingly insignificant decrease in radius will lead to an exponential increase in resistance.

Again, using the traffic analogy, this becomes intuitive. So it's Monday night, and you're heading home from work. You know they're doing construction work on the highway. What do you think would impair traffic flow more, making the construction zone length 2 miles instead of 1, or blocking off three lanes instead of two? That's right! If we block off more lanes, this will result in a major traffic jam as all four lanes attempt to merge down to one. Usually, once you've merged, the traffic picks up the pace a little, right? The flow tends to even out, and it doesn't really matter as much if it's another mile, because at least you're moving! You tend to find that the number of cars tends to get backed up longer and longer prior to the site where the construction begins with everyone tapping their brakes and attempting to merge (the pressure increases before—or upstream from—the narrowing). Conversely, there are far fewer cars on the road

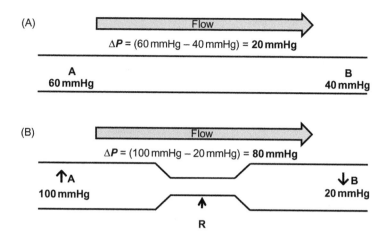

Figure 2.5 Resistance increases pressure upstream but reduces pressure downstream.

after the construction zone ends (the pressure decreases after the narrowing).

The same holds true for fluids! If we look at Figure 2.5, we'll be able to see this same phenomenon with fluids in a glass cylinder. In Figure 2.5A, we see a pressure gradient with limited resistance. If we calculate our ΔP ($60 - 40 = 20$ mmHg), we will see that flow is oriented from point A to point B. However, if we increase the resistance midway between point A and B Figure 2.5B, then we will see three things happen:

1. Flow will decrease overall (remember the inverse relationship between flow and resistance).
2. The pressure upstream of the resistance will increase dramatically (remember construction causes a backup prior to where the merge is).
3. The pressure downstream will decrease (remember once a driver passes the construction zone, there are fewer cars.)

Clinical Correlate

Very small increases in the radius of arterioles will have a dramatic effect on flow. Arteriolar tone is extremely important in maintaining adequate blood flow to each tissue. If all of the body's arterioles dilate at the same time, resistance will decrease so much that pressure in the upstream

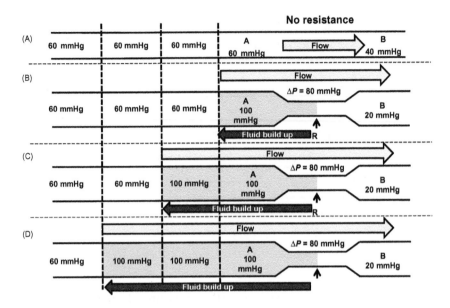

Figure 2.6 Resistance increases pressure upstream because of downstream fluid "buildup."

segments will drop, to the point that oxygen delivery can be compromised (we'll get to this in later chapters). In clinical practice, septic shock and anaphylactic shock are examples of such a situation.

If we "zoom out" a bit, in Figure 2.6, we can see to a fuller extent what happens after narrowing the radius. In these new sections of the cylinder, we can see essentially the brake light/merging effect. We can dub this phenomenon: fluid buildup! (Keep this in mind, as it is an extremely important concept that will come into play a little later.)

2.6 PRESSURE CAN MODIFY RESISTANCE

Bear in mind that the flow equation we described above is for rigid tubes such as glass cylinders under ideal laboratory conditions. In the human body, however, the interplay between pressure and resistance can be a bit more complicated. Ever try and blow up a balloon? Do you notice how it's harder in the beginning and it becomes much easier as you overcome its resistance? Well, blood vessels are not glass cylinders. They are also not exactly balloons. But they do dilate blood vessels and their resistance will somewhat decrease when we increase the pressure inside. We will see later on that with greater driving pressures, in certain situations resistance can decrease. For now just understand the general principles. The flow equation and Poiseuille's law both take certain things for granted. Some of the exceptions to these laws will be discussed in the subsequent chapters, and some are beyond the scope of this text. One such example is the effect gravity has on fluid distribution and handling in the body.

2.7 GRAVITY AND ITS RELATIONSHIP TO PRESSURE

We mentioned in the beginning of this chapter that one of the things your body has overcome with respect to fluid movement is gravity. Having adapted our bodies to this planet, we've learned to actually allow gravity to help us do some of the work. Before we get into exactly how, we first have to learn about the effects gravity has on fluid.

Have you ever tried diving into the deep end of a pool or in the ocean? As you start going down, how do your ears feel? When you dive to a certain depth, your ears feel like they're about to explode! Why? Well, water is heavy. As you dive into the water, more and more water is above you. The pressure you feel is due to that water pushing down on you. Remember, pressure is force per unit area. All mass on earth experiences gravitational acceleration and therefore all mass exerts a force due to gravity. The larger the mass, the greater the force. Gravity is pulling the water in the pool/ocean, and everything else on our planet for that matter, toward the center of the earth. The deeper you dive, the larger the mass of water above you. The larger the mass, the greater the force. The greater the force, the greater the pressure! Note that this is still hydrostatic pressure! Its "container" is the earth around it. Again, this is the type of pressure exerted by all fluid when it's at rest.

But this doesn't just apply to massive volumes of water like in the ocean. Using our cylinder model, we can see this same concept at work (Figure 2.7). Gravity has always been acting on our cylinder, however, as we were looking at a horizontal cylinder, it was less obvious as gravity was applying equal forces to all portions of the cylinder because the amount of water pressing down on any given point between A and B was equal, that is, the hydrostatic pressure at point A and point B was the same (Figure 2.7A). If we "flip" our cylinder to

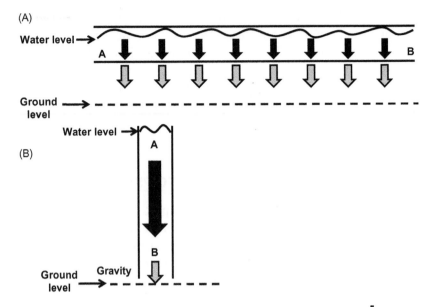

Figure 2.7 The effect of gravity on hydrostatic pressure. (A) in a horizontal tube, gravity ⬇ distributes the amount of water equally, hence the pressure at any point in the tube is equal, because the pressure of the column of water ⬇ is the same at every point. (B) In a vertical cylinder, the column of water above point B will press down making the pressure at point B greater.

an upright position, then what we generate is a phenomenon like the one we described in the "pool diving" example. Now, the pressure at point B is larger than the pressure at point A because the entire weight of the column of water is "pressing down" on point B, whereas in point A, the only thing pressing down on it is air. Now keep in mind, as we said before, water is not compressible. So, much like extra books sitting on a desk, this extra pressure won't necessarily make the water move. However, if there is room to move and there is a pressure gradient, then movement will occur. In the capillaries of the human body, there exists just such a situation, but as we'll see, it involves an integration of all the things we've talked about so far in first two chapters.

2.8 STARLING'S FORCES

In the first chapter, we mentioned that two-thirds of the fluid in your body is intracellular and that, of the one-third that is extracellular, the majority is interstitial. So, plainly stated, an average cell never actually sees or directly communicates with blood! (Figure 2.8) The capillaries exchange nutrients and metabolic waste with the interstitial fluid, which

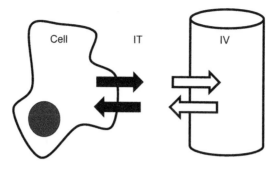

Figure 2.8 Exchange occurs using the interstitial (IT) fluid as a buffer! The capillaries (IV) exchange nutrients and metabolic waste with the IT (white arrows), which in turn will exchange with the cell (black arrows). This generates a buffer zone between cell and IV.

then exchanges goods with the cell! Along these lines, if we consider nutrients and waste passengers in a train, the capillaries are the stations and the blood vessels the trains. The capillary is the gateway through which the magic of life happens. For that reason, in the early 1900s, this guy named Ernest Starling wanted to understand them better. Specifically, he wanted to better understand what dictated flow through it. He described four major forces driving most of the exchange that was happening at the level of the capillary. These forces described whether the fluid in the interstitial space would move in or out of the capillaries.

The cool thing is that we've already seen "behind the curtain" in these first two chapters when it comes to the capillary's magic. Remember in Chapter 1 that we defined osmosis as diffusion of water, and we saw that in order for water to move without gravity or a pressure gradient, there needed to be a concentration gradient. In order for there to be a concentration gradient, there needed to be a membrane which was permeable to water but not solutes. We also discussed that it wasn't the type of solute, (e.g., electrolytes) as much as it was the total number of solutes. Additionally, we mentioned that there are other molecules that have the capacity to attract water, such as glucose and proteins. We also stated that there was a difference between your average cell membrane and the endothelial cell membrane of the capillary wall, because the average cell membrane is permeable to water, but is *not* permeable to electrolytes, glucose, or proteins. In contrast, the endothelium is permeable to water, electrolytes, and glucose; however, it is *not* permeable to proteins. Thus, proteins are pretty much "stuck" in place because, in general terms, no membrane is permeable to proteins!

What Starling found was that in actuality mainly two things determined net fluid flow in/out of the capillaries.

1. Hydrostatic pressure
2. Oncotic pressure.

Hydrostatic pressure, which we just discussed, is the amount of water pressure within a container. In the case of the capillaries, it is the distending pressure that the walls of the capillaries feel from the circulating blood volume. Hydrostatic pressure within the vessels can vary with gravity; however, hydrostatic pressure from fluid flow is always there! (astronauts in space still have hydrostatic pressure driving blood along their blood vessels even though they're weightless!), while oncotic pressure is basically the pressure due to proteins. Now before you get bent out of shape because there is another type of pressure to learn, relax! Oncotic pressure is just the osmotic gradient force due to proteins. The only reason this is given special consideration is because glucose and electrolytes can move in and out of the capillary with water, so there is no osmotic gradient (there is no net "pull" on water) between IT and IV. Proteins, however, can't move between IV and IT and thus can generate pull, that is, oncotic pressure. Albumin, the most abundant protein, is constantly being produced by the liver and secreted to the IV space. Thus albumin is the most important generator of oncotic pressure! The intracellular fluid doesn't change much with changes in oncotic pressure because, among other reasons, it has plenty of its own intracellular proteins.

Key

As long as there is no net change in total EC osmolality, there will be NO fluid shift between the EC and the IC compartments.

Analyzed from another point of view: electrolytes freely distribute in the Extracellular Fluid (ECF); however, they can't cross the cell membrane, proteins can't cross the capillary membrane which means that they will not be able to diffuse into the interstitial space and thus are confined to the IV space.

Ok, so let's look at Starling forces in a bit more detail. If you look at Figure 2.9, you'll see that these Starling forces work both ways. We have capillary hydrostatic pressure (C_H) and it forces fluid out of the

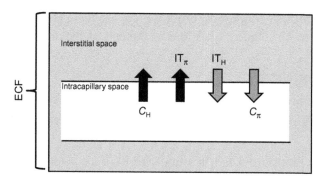

Figure 2.9 Starling's forces. The forces that favor water movement from the intracapillary space into the interstitial space ⬆ are the capillary hydrostatic pressure and the interstitial colloid-osmotic pressure. The forces that favor water movement from the interstitium to the intracapillary space ⬇ are the interstitial hydrostatic pressure and the capillary colloid-osmotic pressure. C_H is the intracapillary hydrostatic pressure, C_π is the intracapillary colloid-osmotic pressure, IT_H is the Interstitial hydrostatic pressure, IT_π is the interstitial colloid-osmotic pressure.

blood vessel. Conversely, any liquid that is located in the interstitial space will also exert pressure on the vessel wall, but in the opposite direction. This is known as interstitial hydrostatic pressure (IT_H).

> **Key**
>
> Intracapillary hydrostatic pressure forces fluid out of the blood vessel.
> Interstitial hydrostatic pressure forces fluid into the blood vessel.

This means that both the C_H and the IT_H will oppose each other, and fluid shift will follow the greatest pressure. In general terms, if C_H is greater than IT_H, fluid will leak out of the capillaries and if the IT_H is greater than the C_H, the fluid will shift into the capillaries. The second force that determines fluid shift between the capillaries and the interstitium is the oncotic pressure (also known as colloid-osmotic pressure). Oncotic pressure is generated by the free proteins that are located in the extracellular fluid. There tends to be a lot more free proteins in the intravascular space than in the interstitial space. As proteins are not permeable to the capillary wall, they will exert an osmotic force on water, which IS permeable. This means that the intracapillary oncotic pressure C_π will tend to keep water into the capillaries and the IT_π will tend to keep water in the interstitium. Changes in any of these forces will alter the distribution of fluid! If plasma oncotic pressure decreases, then the capillaries will no longer be able to retain water. This will have tremendous repercussion clinically, as we see in congestive heart failure, cirrhosis, and nephrotic syndrome. If you're not yet familiar with these conditions, don't worry! We'll explain them in a little more detail in our clinical questions at the end of the chapter.

Clinical Correlate

A decrease in plasma oncotic pressure will favor fluid leak into the interstitial space generating edema!

The key points to keep in mind are that the capillary hydrostatic pressure, C_H, is going to be directly related to both the ΔP (driving pressure) that was discussed earlier, and the resistance of the blood vessels both before and after our capillary. Why do Starling's forces only affect the capillaries? Because it is only at the level of the capillary wall (which is permeable to liquids and solute, but NOT permeable to proteins) that an exchange can occur between the intracapillary and the interstitial fluids.

In this chapter, we have analyzed how fluid moves within the body and some of the major forces that are associated that are behind fluid movement in the capillaries. In the next chapter, we will analyze how the heart and blood vessels use the concepts of resistance and pressure to shuttle blood around the body.

2.9 CLINICAL VIGNETTES

Example 1

A soldier was reprimanded and ordered to stand-at-attention for a full 12 h shift in front of the entrance to the military base. After which he arrives to the medical office complaining of severe pain and swelling in both of his legs. The on-call physician prescribes some OTC painkillers and recommends elevation of both legs during the night. The change in which of the following was the cause of the soldiers condition?

A. Increased capillary colloid-osmotic pressure
B. Decreased capillary colloid-osmotic pressure
C. Increased capillary hydrostatic pressure
D. Decreased capillary hydrostatic pressure
E. Increased interstitial colloid-osmotic pressure

F. Decreased interstitial colloid-osmotic pressure
G. Increased interstitial hydrostatic pressure
H. Decreased interstitial hydrostatic pressure.

Answer C. The soldier was forced to maintain a static position for 12 h thus increasing the hydrostatic pressure within the veins and arteries of the lower limbs. (Remember gravity exerts additional hydrostatic pressure.) By not being able to move and therefore inactivating the venous muscle pump, there was fluid accumulation within the capillaries, thus there was an increase in capillary hydrostatic pressure, which favored fluid leakage into the interstitial space (hence the edema).

Example 2

A 7-year-old boy arrives to the emergency department with a 1 month history of generalized edema. It started in his extremities but has progressively gotten worse. He has gained 10 lb in the past 2 weeks. His mother says he's been urinating copiously and it appears "foamy." On arrival to the office, you notice 2–3 + pitting edema of all extremities, scattered crackles in his lungs and upon direct questioning a history of a viral prodrome about 1 week before the onset of symptoms. The element of Starling's forces which is altered in this patient is:

A. Increased capillary colloid-osmotic pressure
B. Decreased capillary colloid-osmotic pressure
C. Increased capillary hydrostatic pressure
D. Decreased capillary hydrostatic pressure
E. Increased interstitial colloid-osmotic pressure
F. Decreased interstitial colloid-osmotic pressure
G. Increased interstitial hydrostatic pressure
H. Decreased interstitial hydrostatic pressure.

Answer F. This patient is suffering from nephrotic syndrome, which is characterized by more than 3 g of protein in the urine in 24 h. This loss of intracapillary oncotic pressure will favor extravasation of fluid into the interstitial space and thus generating edema. Conditions associated to decreased oncotic pressure (nephrotic syndrome, cirrhosis, malnutrition) can be associated to edema because the decreased oncotic pressure will favor leakage into the interstitial space.

Example 3

A 24-year-old football player breaks his right wrist during training. He is immediately brought to the ED where the bone is reset. The doctor instructs him to keep his arm elevated to avoid pain. The football player forgets all the doctor's instructions and begins to feel an excruciating amount of pain in right arm that is relieved when he elevates it. The pain is likely generated by which of the following:

A. Increased capillary colloid-osmotic pressure
B. Decreased capillary colloid-osmotic pressure
C. Increased capillary hydrostatic pressure
D. Decreased capillary hydrostatic pressure
E. Increased interstitial colloid-osmotic pressure
F. Decreased interstitial colloid-osmotic pressure
G. Increased interstitial hydrostatic pressure
H. Decreased interstitial hydrostatic pressure.

Answer G. The inflammation that results from the fracture leads to increased delivery of blood flow and subsequently because of increased capillary permeability which is also favored by inflammation, there is an increased extravasation of fluid into the interstitial space. (Remember the cardinal signs of inflammation!) When the football player has his arm swinging at his sides, the hydrostatic pressure increases within the capillaries (this however happens to all of us and we don't scream and shout in pain). In this scenario, however, the increased amount of leakage leads to increased interstitial hydrostatic pressure which leads to edema, if on top of that we add the effect gravity can have on fluid distribution, this will lead to even more swelling and consequently pain!

CHAPTER 3

Fluid Movement in the Body: Primer to the Cardiovascular System

3.1 FLUID HANDLING BY THE HEART AND CARDIOVASCULAR SYSTEM

So far we have discussed fluids in isolated systems. We will now try to integrate the independent concepts that were presented with a fluid set of ideas (pun intended). The reader will notice that both the heart and the blood vessels adapt to each other, this means that they will change in order to maintain blood flow during different circumstances (decreased circulating volume, decreased pressure, decreased heart function among other things), with various simultaneous responses. A thorough explanation of each situation is beyond the scope of this book. However, the main focus of this book provides the basic knowledge behind the integration of how distribution, movement, and regulation of fluid plays a key role in the transport of oxygen to the tissues. Therefore, we will discuss the adaptive responses that allow the maintenance of blood flow, blood pressure, and oxygenation. However, for an in-depth discussion of any of these topics, the reader should refer his/her favorite physiology textbook.

3.2 RESISTANCE AND FLOW ALONG THE CV SYSTEM

The cardiovascular (CV) system includes the heart and the blood vessels. Remember from our discussion in Chapter 1 that the fluid within the CV system (plasmatic fluid) represents approximately 1/3 of the ECF.

Key

The other 2/3s of the ECF are in the interstitial space, BUT THEY SHARE THE SAME CONCENTRATIONS OF ELECTROLYTES AND WATER!

The intravascular fluid within the body is distributed along the CV system. There are two pumps that move the fluid around: the heart and the venous muscle pump. It is through the concerted action of these two pumps that blood will flow around the CV system.

The CV system is a complex system of pressure and flow; however, in order to simplify understanding of flow and resistance along the entire CV system, we will depict it as a single tube with sequential points of resistance along the flow of blood (Figure 3.1). In this schematic, we have divided the right side and the left side of the heart, so that it is clear where the blood is going. Thus, the blood will start in the LV and will encounter the first point of resistance R1 in the systemic capillaries, it will continue through the venous system toward the right ventricle (R2), then onward to the lungs and the pulmonary capillaries (R3) finally to end up in the left ventricle (R4). Even though the heart is effectively the pump that pushes fluid through the whole system, describing it at least initially as a point of resistance can make the flow throughout the CV system a little more intuitive, in that any alteration in pump function we can equate with an increase in resistance across that particular point, and an increased function can be equated to a decrease in resistance. So, if we see the CV system as a closed circuit with four points of resistance placed in series (Figure 3.1), the exact same principles that were outlined in the previous chapter with regard to flow: Flow $= \Delta P/R$ hold true! This means

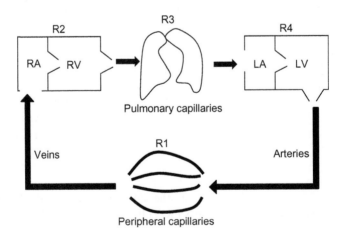

Figure 3.1 Points of resistance to flow along the systemic and pulmonary circulations. As blood flows from the LV, through the systemic circulation, and back to the RV and the lungs, it will pass four points of resistance that will determine the hydrostatic pressures in the upstream and downstream segments.

that an increase in resistance, will increase pressure upstream and decrease pressure downstream!

Clinical Correlates

Pharmacological modification of R1, arterioles, can have a huge impact on blood pressure. An infusion of phenylephrine (a potent vasoconstrictor) increases R1 and thus increases blood pressure. Phenylephrine (Levophed®) is used as treatment in some patients who can no longer maintain their BP!

Looking at Figure 3.1, we can see how a change in the R at any one of our four points of resistance in the system **will affect the entire body!** For example, if we generate arteriolar vasoconstriction (increase R1), the pressure will increase upstream (arterial system)!

The take-home message from Figure 3.1 is that the vascular system as a whole is affected by each and every point of resistance. The pressure gradients generated both upstream and downstream of these different points of resistance will affect the pressure flow dynamics of how blood circulates throughout the body. This means that in order to maintain constant flow throughout the system, the same amount of blood has to flow through each segment! Let us say that again, the same amount of blood has to be going through each segment of the CV system at any given point in time. This does not mean that the total amount of blood in each portion of the vascular system is the same (at any given moment veins have the largest amount of blood in any other system). What it means is that, in order for pressures in each segment to remain stable, there has to be an equal amount of flow BETWEEN segments. This means that the same amount of blood that travels across the RV has to pass through the lungs (otherwise blood would pool in the RV), and that same amount of blood that just passed through the lungs has to be ejected by the LV (or blood would build up in the lungs). To clarify this point, imagine that you're trying to fill a tub with water. If you open the faucet and there is a flow of 1 L/min and the drain is properly sealed, water will begin to accumulate in the tub. Imagine that after 30 L of water, you open the drain and there is an outward flow of 1 L/min. If the faucet is still open, then the total amount of water in the tub will remain unchanged, the flow across the faucet→tub→drain is the same (1 L/min) even though the amount of water in the faucet, tub or drain is different! This is similar to what happens with the vascular system, in that flow is the same throughout the system even though the total amount of blood in different segments may vary widely. If you alter either the resistance or flow from faucet (turning it on or off) or the drain (clogging it with a glob of hair), then the amount of water in the tub will change.

The CV is very similar to our tub example in that

1. Flow is equal in all portions of the CV.
2. Total volume can vary along the CV.
3. By altering resistance we can alter both flow AND volume.

Now that we know how the CV is interconnected let's look at the parts of our system separately.

3.3 THE BLOOD VESSELS

How exactly do blood vessels control resistance and flow? So far in Chapter 2, we reviewed the effects of increasing resistance (vasoconstriction) or decreasing resistance (vasodilation), and the effect this has on flow. We have two different systems that carry blood in the body: the arterial system and the venous system, each with its own set of peculiarities.

The main function of arteries is to carry oxygenated blood from the left side of the heart to the rest of the body. As such, they need to keep pressure inside so that flow is possible. The thick smooth muscle layer provides the arteries with the capacity to hold large pressures but low volumes (arteries have approximately 1/3 of the circulating volume). The main function of veins on the other hand is to return blood to the heart and lungs for oxygenation, which means that veins handle large amounts of volume (approximately 2/3 of the circulating volume) but low pressures! Characteristically veins have a very thin smooth muscle layer that will allow them to contract and dilate, but all the while keeping pressure low. Therefore, we can say that compliance, which is the amount of pressure in a container given a change in volume, is different for arteries than for veins. Arteries have low compliance (low volume, high pressure), while veins have high compliance (high volume, low pressure).

Key

Compliance is defined as the change in pressure that occurs after a change in volume has taken place, the formula is: $\Delta Vol/\Delta Pressure$. This means that the more compliant something is, the less the pressure will change after a given volume and vice versa, a compliant vessel (veins) will accommodate large amounts of volume with modest increases in pressure, while in a noncompliant vessel (arteries) even a small increase in volume will increase the pressure greatly.

Putting this together, remember the tub example from the previous section. The tub has a total volume of 30 L with the faucet open at a flow rate of 1 L/min and a drain flow rate of 1 L/min. The total volume of the tub IS NOT CHANGING; however, **there is** water flowing along the tub. What would the pressure in the tub be? There would be low pressure because the 30 L is comfortably placed within the tub. Now imagine that the drain has a 2 m long 1 in. pipe connected to it. This pipe has to accommodate a flow of 1 L/min (same as the tub) but in a much narrower space, therefore pressure is increased in this segment, in order to keep flow constant. Veins and arteries work in a similar fashion in that flow is the same throughout the circulatory system but the volumes are different between segments, therefore pressure in the arterial system is high while pressure in the venous system is low. During periods of stress, fluid can be redistributed between segments (veins to arteries) in order to favor increasing blood pressure.

3.3.1 Venous Muscle Pump

Before we discuss the heart, the body's main pump, let's briefly analyze the venous muscle pump system. Gravity exerts a positive pressure on blood flow from the head to the toes in an upright individual, that is, as discussed previously by "pulling" the water toward the earth, this increases the hydrostatic pressure within that vessel. If we look at Figure 3.2, in relation to venous blood flow, we can regard the heart as the equilibrium point as all blood will tend to flow toward it. This means that gravity will favor venous drainage of blood in tissues that are above the heart and will oppose venous flow in areas that are below the heart. (The arterial system is designed so that that the driving pressure generated by the heart is enough to overcome gravity, so arterial blood flow is relatively independent of gravity.) So back to the veins, if venous blood flow from the lower limbs to the head is

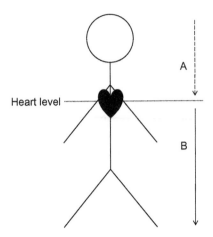

Figure 3.2 The effect of gravity on venous blood flow. Considering the heart, the equilibrium point in our system (all venous blood flow MUST return to it), gravity will favor blood flow from body areas higher than the heart (A), and oppose flow in areas that are lower than the heart (B). In an upright individual, this generates positive pressure in the legs and negative pressure in the head and shoulders.

opposed by gravity, how is it that the body can overcome this and manage to get blood back to the heart? Figure 3.3 shows us exactly how the venous muscle pump works. The key components are (1) the venous valves and (2) the muscles themselves. Valves work like upside down parachutes, when blood flows distally (i.e., toward the hands/feet), the valves will fill up with blood and close (this increases the R toward infinity and thus stops flow). However, when blood flows toward the heart, the blood venous valves will open (thus R will decrease and flow will increase!). In order for blood to flow, a pressure gradient must be generated. By contracting, muscles will compress the venous walls and thus will generate an increase in hydrostatic pressure inside the veins (because their container is shrinking), favoring flow in the direction in which the valves open up!

Clinical Correlate

Venous insufficiency is characterized by nonfunctioning valves that allow flow in the opposite direction! This will paradoxically increase the hydrostatic pressure in the downstream segments of the vein (as blood accumulates because of valve dysfunction, this increases the intracapillary hydrostatic pressure, favoring fluid leakage into the interstitial space). An increase in interstitial fluid will increase the oxygen diffusing distance. If very severe, this can lead to tissue ischemia over time, and the clinical signs and symptoms associated with venous insufficiency.

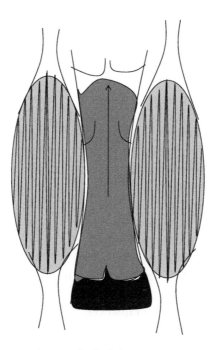

Figure 3.3 As the muscles compress the vein walls, the hydrostatic pressure within a vein segment will increase, favoring blood flow toward the path of least resistance, resembled by the open valves vs. the closed ones.

3.4 THE HEART

The heart, however, functions differently than the blood vessels in how it propels blood along the CV system. We have labeled the entire right side of the heart as R2 and the left side of the heart as R4, because conceptually any decrease in ventricular function can be interpreted as an increased resistance. The heart however does not act in the same way as a blood vessel point of resistance because it can't readily modify its radius as blood vessels can. So, why do we label the heart as two different points of resistance? Resistance was defined as opposition to flow (Chapter 2). In that case, then the heart does provide an opposition to flow! If the heart is not pumping blood adequately, resistance to flow increases, that is, blood won't go through the atrium/ventricle. So for now, let's consider for now that each side of the heart is a fixed point of resistance.

In truth numerous factors come into play with regard to ventricle function, far more than can be described in this textbook; however, we will try to emphasize what we consider to be the most important concepts behind ventricular function.

3.4.1 The Cardiomyocyte

In order to discuss the heart, a brief description of the characteristics of striated muscle is warranted. The functional unit of the muscle is the sarcomere (Figure 3.4A,B). Each sarcomere (i.e., muscle unit) is composed of actin and myosin filaments which overlap, and thus can pull the A bands (where each sarcomere ends) closer together, thus reducing sarcomere length. The molecular basis of muscle contraction is a calcium-dependent process, which will generate a conformational shift in the actin and myosin proteins, favoring the cyclical coupling and uncoupling of each other. This coupling/uncoupling will generate contraction of the muscle. The force of contraction however will depend on the initial state of interaction between actin and myosin. In Figure 3.4C, the initial interaction between actin and myosin (hashed lines) is close to 50%, this leaves a 50% contractile capacity. When we stretch the muscle as in Figure 3.4D, we see that this allows for a larger, roughly 75% contractile capacity. This generates a larger force of contraction than that of a 50% initial interaction Figure 3.4C. However, if the distance between A-lines increases to the point where the initial interaction between actin and myosin is lost (you stretch the muscle so much the fibers break), then no contraction can occur (Figure 3.4E). To picture this mechanism, imagine

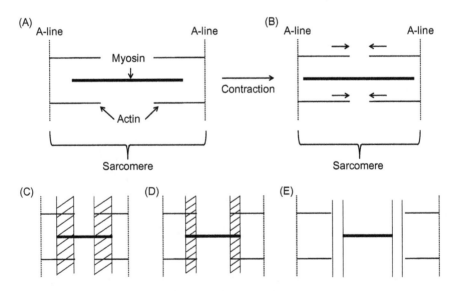

Figure 3.4 During the contraction process the cyclical coupling and uncoupling of the myosin and actin filaments reduces the distance between A-lines (A and B). The progressive lengthening of the sarcomere alters the amount of overlap between the myosin and the actin filaments, and thus the overall strength of contraction (C vs D) until actin and myosin no longer interact and can no longer contract (E).

that you're standing on top of a 6 ft wall and you're trying to pull some-one up. If you and that person hold on from each other's forearm, pulling that person up is easier than if you were say holding on by the hands or finger tips or if you're just yelling at them. . ...

3.4.2 The Frank–Starling Mechanism

These relationships describe skeletal muscle and they hold true as well for cardiac muscle. So, we could say that a muscle resembles a rubber band. The more we stretch the rubber band, that is, the more we stretch the sarcomere, the harder it will contract once we let go, that is until we pull so hard that we break it. This is called the Frank–Starling mechanism of the heart. The more we stretch the heart with the blood that arrives to the ventricle from the atria, while still maintaining an adequate interaction between actin and myosin the harder the subsequent contraction will be. However, if we stretch the heart to the point that we dissociate the actin and myosin filaments completely, the heart will no longer be able to contract (Figure 3.4). So, let's keep this in mind as we move forward.

3.5 THE CARDIAC CYCLE

How does the Frank–Starling mechanism of the heart work to favor contraction? We previously discussed stretching the rubber band and the subsequent force of contraction of that rubber band. The heart works in a very similar manner. The more we stretch the muscle, the harder it will contract! And how exactly do we stretch the heart? The blood that is located in the ventricle BEFORE it contracts will deter-mine how much the ventricle stretches and will determine the length of the sarcomere. Therefore, analyzing these relationships becomes a lot simpler if we look at the pressure–volume curve of the heart (Figure 3.5). This figure is a representation of the relationship between volume and pressure in the left ventricle along the cardiac cycle (the same curve holds true for the right ventricle, although the pressures are much lower). Let's start at the beginning of diastole. At point (a) the ventricle has just finished squeezing out the majority of the blood con-tained within. We can see that there are approximately 50 mL of blood in the left ventricle. Because the ventricle is relaxed, the pressure exerted by that small volume is nearly 0 mmHg. However, as the clo-sure of the AV valves has dramatically increased resistance to blood traveling from the atrium to the ventricle, this causes a "pooling" of

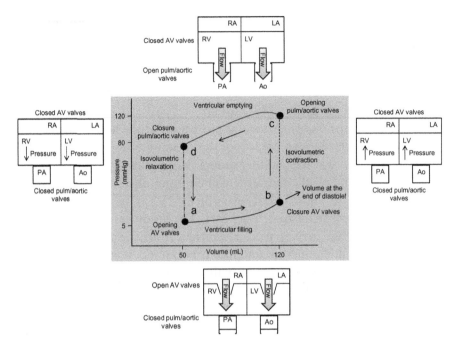

Figure 3.5 Left ventricular volume—pressure curve (see text for details). AV, atrioventricular; RA, right atrium; LA, left atrium; RV, right ventricle; LV, left ventricle; PA, pulmonary artery; Ao, aorta.

blood in the atrium. This "pooling" of blood then leads a buildup in hydrostatic pressure. This buildup in pressure in turn leads to an increase in the pressure gradient between the atrium and the ventricle. Eventually this gradient becomes so high that it overcomes the AV resistance and pushes open the AV valves, filling the ventricles with blood. We can see, however, that even though the blood volume is increasing along segment (**a**—**b**) the pressure only rises modestly, this means that the ventricles fill up with blood, but there is only a small increase in pressure. This is because the AV valves are open and the myocardial muscle is relaxed. Thus no pressure can be generated within the ventricular cavity. (Imagine trying to inflate a balloon with a hole in it.) This changes abruptly when ventricular contraction starts. As the muscle begins to contract, this will rapidly increase the pressure inside the ventricles and force the AV valves shut. (Imagine the AV valves are a parachute and blood is the air, when the blood attempts to flow from the ventricle to the atrium, they fill up with blood thus preventing backward flow. When the blood flows from the atria to ventricles it forces the "parachute" shut, allowing flow to occur.) Once the AV valves are shut, the pressure starts to increase rapidly, a phase known as isovolumetric contraction, segment (b–c). (It's dubbed isovolumetric because the volume doesn't change! Iso = equal.) The pressure increases because the aortic and pulmonary valves are shut and the ventricles are beginning to squeeze. (There is more pressure in the aorta and pulmonary artery than in the ventricle, this forces the aortic and pulmonary valves shut, until the pressure in the ventricles is greater than the pressure in the aorta and pulmonary artery.) The pressure inside the ventricle will continue to increase until it overcomes the pressure in the aorta/pulmonary artery at which point the semilunar valves will open (c) and the ventricle (which is still contracting) will empty its contents into the aorta and pulmonary artery. This emptying will continue until the ventricles lack enough blood to maintain the pressure needed for the semilunar valves to stay open (d). At this point, the ventricle will begin to relax (isovolumetric relaxation). Even though the pressure in the ventricle is no longer high enough to maintain the semilunar valves open, it is still higher than the pressure in the atrium and thus only until the pressure drops below the pressure in the atria, will the AV valves open (a).

Clinical Correlate

A premature ventricular contraction (PVC) is an asynchronous beat from the heart, which is premature and thus can only achieve partial emptying of blood into the aorta. This will in turn increase the end systolic volume (ESV). Because venous return stays constant all throughout the PVC, the LV will end up with more blood than it's supposed to have, that is, the blood arriving from the pulmonary veins and the "extra" blood left over from the PVC. This increased ESV will stretch the walls of the heart (think rubber band), thus generating a more forceful contraction, which will displace the blood that has accumulated, returning ESV to baseline.

3.6 QUANTIFYING HEART FUNCTION

In order to quantify how much blood the heart is actually pumping, we must understand three particular concepts: stroke volume, ejection fraction, and cardiac output.

Stroke volume—The amount of blood that is expelled from the heart with each contraction. This is easily calculated by looking at Figure 3.5. We mentioned that the volume at point (b) is the volume that "resides" in the left ventricle right before it begins to contract (approximately 120 mL) also known as end diastolic volume (EDV), and point (a) is the volume that "resides" in the left ventricle after systole has occurred (approximately 50 mL). So, let's substract the residual volume (a) of 50 mL from the initial volume of 120 mL, and we get a 70 mL difference. This means that the ventricle is ejecting approximately 70 mL of blood per beat. This is called the **stroke volume** or **SV**, and the formula is: (stroke volume (SV) = end diastolic volume (EDV)−end systolic volume (ESV)).

Ejection fraction—How efficient is the ventricle at pumping blood? The **ejection fraction** or **EF** is an indicator of how efficient the ventricle is at emptying itself. It's the percentage of the EDV that is ejected from the ventricle. The formula is: EF = SV/EDV. (If we want to turn this into a percentage, we simply multiply by 100). A normal ejection fraction is above 60%. This means that the ventricle is able to efficiently eject 60% of the EDV.

Clinical Correlate

A decrease in the EF means that the ventricle is no longer able to contract efficiently, and thus blood will accumulate in the immediate upstream segment, if it's the LV, blood will accumulate in the lung capillaries, if it's the RV, blood will accumulate in the jugular veins, liver, and lower extremities.

Cardiac output—The amount of blood that the heart pumps per minute! We stated previously that the stroke volume is the amount of blood the heart ejects per beat. If we multiply the stroke volume (SV) × heart rate (HR), we will get the total amount of blood that is pumped by the heart per minute, this is known as **cardiac output** or **CO**. The formula is: CO = SV × HR. This is approximately 5 L/min, SV = 70 mL, HR = 70 BPM → 70 mL × 70 BPM = 4900 mL/min. Cardiac output is an extremely important number; it serves as an indicator of how the heart is functioning with respect to tissue demand of oxygen. Wait, what? We will go into this a little bit later on, but the CO represents the amount of blood that the heart is delivering to the tissues, and if the blood's main job is to deliver O_2, then CO can be correlated (in most cases) to O_2 delivery to the tissues. So, if the tissue demand for O_2 increases, then CO should increase accordingly. Along these lines there is a formula, which is KEY to understanding how CO and oxygen interact.

$$\text{Oxygen delivery} = CO \times \text{oxygen carried by hemoglobin}$$

The first part of the formula is CO, therefore increases in CO increase O_2 delivery and decreases in CO decrease O_2 delivery. Now, let's understand how we can modify CO.

3.7 PRELOAD AND AFTERLOAD

Let's understand how the Frank–Starling mechanism is related to the cardiac cycle and CO. If we look closely at point **b** in Figure 3.5, we can see that this represents the volume inside the ventricle at the end of diastole. Imagine a water balloon connected to a faucet and then turning the faucet on. The more water flows inside the balloon, the greater the balloon will stretch. The exact same thing happens with the ventricles. The more blood is in the ventricle at the end of diastole, the more the ventricle wall is being stretched, and thus the harder contraction will be that is until the uncoupling of actin and myosin takes place (see above). The length of the cardiac muscle cell at the end of diastole which is determined by the left ventricular end diastolic volume or LVEDV is what we call preload! Preload, or the load before the contraction, is determined by the amount of blood that gets to the ventricle during diastole, how much the muscle stretches, and hence the amount of blood that the ventricle has to contract against. As we defined earlier on that the same amount of blood has to circulate through the circuit at all times, then we can extrapolate this concept to say that both the LV and the RV have to pump the same amount of blood, otherwise blood would be accumulating either in the systemic venous end or in the lungs. Remember that flow is equal across the CV system! Therefore, we can state that venous return is a great indicator of preload! Why? If venous return is the amount of blood getting to the right side of the heart and both sides of the heart have to pump the same amount of blood in order to avoid the accumulation of fluid, then the LV has to pump whatever the veins are returning to the heart and this is equal to the CO! This means that the greater the venous return, the greater the preload, the stronger the contraction and thus the greater the CO! Figure 3.6A is a graph of exactly this phenomenon. The x-axis represents preload, this is measured indirectly as LA

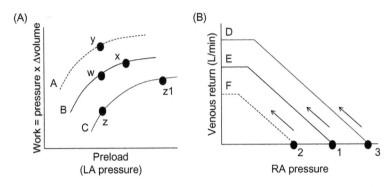

Figure 3.6 Cardiac and vascular function curves. (A) Cardiac function will vary according to preload and contractility. A change in preload will move along the same line (w→x); however, a change in contractility will change the response of the myocytes to a given preload (w→y; w→z). (B) Venous function curve is a direct function of MSFP and cardiac function which decreases RA pressure (see text for details).

pressure. Why indirectly? As we can't measure the length of the cardiac muscle before contraction, we use pressure to indicate how much stretching is actually going on inside. (Remember: If there is a lot of fluid in the LA, then the pressure will increase; thus, the more the volume, the higher the pressure.) The y-axis represents the work that is exerted by the LV. The formula represented is Work = Pressure × ΔVolume. Roughly translated into layman's terms it means how much work, in the form of pressure, does it take to eject a specific amount of blood (Δvol = SV). If we look at line B, we can see that the higher the LA pressure (preload), the more work is going to be carried out by the left ventricle! This is a graphical representation of what we've been studying the past couple of pages, that is, the Frank–Starling mechanism.

Clinical Correlate

Central venous pressure (CVP) and pulmonary capillary wedge pressure (PCWP) are both measures of cardiac function. CVP is measured in the atriocaval junction (right where the cavas (IVC and SVC) empty into the RA), and it's used as an indicator of preload. The more volume there is in the RA and RV, the higher the pressure in the venous system will be. PCWP is obtained via a Swan–Ganz catheter. This catheter is threaded through the right side of the heart until it's positioned inside a branch of the pulmonary artery, thus indirectly measuring the pressure of the LA. Both of these methods are used to estimate the EDVs of the RV and LV; however, they're only estimates!! Why? CVP and PCWP can be modified by a great number of factors other than volume itself, one of which being the compliance of the ventricle itself. If the ventricle wall is stiff (low compliance), a small amount of volume will generate large increases in CVP or PCWP, which does not necessarily mean that the patient is volume overloaded, thus these measurements should ALWAYS be interpreted on a patient-to-patient basis.

What is afterload? Afterload is the amount of pressure that the ventricle needs to overcome in order to empty its contents into the aorta. We defined preload as a volume-dependent phenomenon, that is, preload = LVEDV; however, afterload is defined in terms of pressure! The more pressure there is in the arterial system, the harder it's going to be for the LV to eject blood into the aorta. Let's analyze this concept a little bit further. What determines afterload? Remember our discussion from Chapter 2: increase R and you increase the pressure upstream. Well, arterioles are the first point of resistance that is immediately downstream of the LV. By increasing R at the arterioles, pressure increases in the aorta, that is, increasing afterload. So, afterload is a direct result of total peripheral resistance or TPR (which is the sum of all the Rs in the capillaries). If TPR increases, for example, from NE-mediated vasoconstriction, this will automatically increase the pressure in the arteries and this will increase afterload.

Key

Assuming there are no changes in circulating blood volume, TPR determines afterload.

So, if we increase the pressure in the arteries, we increase afterload. How does the heart compensate for increased afterload? During a normal cycle the LVEDV will be approximately 120 mL, and the LV will eject 70 mL, leaving 50 mL inside the LV at the end of systole (ESV). When the mitral valve opens, another 70 mL will flow into the LV and the LVEDV will be 120 and the cycle starts all over again. Well, imagine that there's an acute (min) increase in afterload. This will immediately hinder the LV's capacity to eject blood, and it will not be able to have an adequate stroke volume (in this case the LV will eject 50 mL). This will pool blood in the LV at the end of systole, increasing the ESV from 50 to 70 mL. Venous return however is still normal, so the amount of blood that is entering the LV from the LA is still 70 mL.

The problem is that instead of there being 50 mL in the LV, there is 70 mL. This will increase the LVEDV to 140 mL! Remember the Frank–Starling mechanism? This increase in LVEDV will stretch the heart further, favoring a stronger contraction that will theoretically be able to overcome the increased afterload. So, the cycle will start again, but the new set point for LVEDV or preload will be 140 mL instead of 120 mL.

Clinical Correlate

Patients with congestive heart failure have lost the capacity to adapt to increased afterload. It is therefore key to maintain the arterial blood pressure controlled, as patients can go into acute decompensated heart failure if afterload increases and the diseased heart is not able to compensate!

If we look at Figure 3.6A, line B is represented by the movement from point W to point X. Generally, any sudden increase in TPR (and thus afterload) will be mediated by hormones that also increase ventricular function so, the increases in LVEDV are hardly noticeable, for example, NE and E. However, a persistent increase in TPR (be it mechanic, e.g., aortic stenosis, or hormonal, e.g., hypertension) will strain the LV. In response, the LV will hypertrophy (just as a weight lifter does after prolonged exercise) in order to overcome the increased afterload. This is detrimental over time because, among other things, it increases the myocardial demand for oxygen, it decreases coronary vessel function and it decreases ventricular filling during diastole, all of which could lead to a myocardial infarction or heart failure.

Clinical Correlate

Mean arterial pressure (MAP) is a number that is used to determine perfusion. If we have an adequate MAP, this means that perfusion is maintained to certain key organs. (Vasoconstriction is selective, based on the expression of receptors along the vessel wall, arteries that go to the brain and heart have a comparatively decreased expression of alpha-chatecolamine receptors and as such are MUCH LESS susceptible to NE-induced vasoconstriction.) The formula to calculate MAP is the following: $(1/3 \times \text{Systolic Pressure}) + (2/3 \times \text{Diastolic Pressure})$. This is because the systolic pressure corresponds to 1/3 of the cardiac cycle, and the diastolic pressure corresponds to the other 2/3 of the cardiac cycle.

Along with the Frank–Starling mechanism of the heart and the overlap between actin and myosin filaments, there is another characteristic that we need to mention before moving on, we call that contractility. Contractility can be defined as any increase in the force of contraction (Work) that CANNOT be attributed to the Frank–Starling mechanism of the heart. This means that for the same preload (stretching) of the heart muscle, there is a stronger contraction. Let's look at Figure 3.6A again. If we look at line B, we can see that the amount of work increases or decreases based on preload. However, look at line A. It has more or less the same form as line B, but it shows more work being done at the SAME PRELOAD. More specifically, look at points W and Y. Both have the same preload however at point Y the heart is carrying out much more work. This is called contractility or how "responsive" the cardiac muscle is.

Clinical Correlate

In Figure 3.6A, there are three curves: A, B, and C. B represents the contractility of a normal heart, while A represents a heart with increased contractility. Curve C represents a heart with DECREASED contractility. A heart that has suffered a previous infarction and has a noncompliant ventricular wall, a heart with dilated cardiomyopathy (due to alcohol or a virus) would be examples of hearts with decreased contractility. A heart with decreased contractility is much less responsive at a given preload, that is, it will exert LESS work at a given preload (point Z). This means that a much larger preload has to be achieved in order to carry out the same amount of work as a normal heart (point Z1). This increased preload increases the hydrostatic pressure in the downstream segments (remember a failing heart represents and increased R) and patients with heart failure have a much higher risk of developing pulmonary edema.

How exactly does this work? Well, a simple explanation (albeit not complete) could be attributed to the amount of calcium, and thus the speed of the coupling/uncoupling of the actin and myosin filaments which leads to a more forceful contraction. Anything that increases the intracellular concentration of calcium will favor a faster coupling/uncoupling of the actin and myosin filaments, and thus will increase the strength of contraction. The activation of beta-receptors by catecholamines is one such mechanism of increasing the force of contraction by increasing contractility.

Clinical Correlate

The next time someone scares you, feel your heart. You will not only feel it beating faster, but you will feel it beating harder! This increase in the force of contraction is mediated by the catecholamines that were secreted when you were scared!.

Continuing with our rubber band example, the rubber band that we discussed previously was a 0.5 mm wide rubber band, if we were to "increase" the contractility of our rubber band, it would be a 1 mm or even 2 mm wide rubber band. This means that the force of contraction would be even greater in the second, wider rubber band. Translated into practical terms, this means that there are essentially two mechanisms through which we can increase the relative force of contraction: (1) Frank–Starling mechanism of the heart and (2) contractility. The Frank–Starling mechanism is affected by how much the heart's muscle fibers are stretched before it begins to contract; this is where the concept of preload comes in. Preload is the amount of blood that is in the ventricle at the end of diastole, it is also known as the VEDV. VEDV (right or left) will stretch the wall of the ventricle thereby changing the way the actin and myosin overlap, thus the larger the preload, the stronger the muscle will contract, up to the point where the dissociation of actin and myosin occurs (Figure 3.2E). Preload is determined by venous return, that is, the amount of blood returning to the heart. If we return to the concept previously introduced in this chapter in which the flow through the entire CV system is the same, then we can assume that the venous return will be equal to the amount of blood leaving the heart. This is why we consider ventricular volumes to be relatively similar between the right and the left ventricles.

3.8 VASCULAR FUNCTION CURVE

What determines preload? In Figure 3.6B, we see a venous function curve. The whole point of the venous system is to return blood to the heart so that it can pump it into the lungs and back into the arterial circulation in order to provide oxygen. We said previously that the amount circulating through the circulatory system, that is, the blood passing through any section of the circulatory system, is the same (otherwise there would be pooling of blood). So, the amount that flows through the venous system has to be the same amount that is flowing through the arterial system, the lungs, and even the heart. How is this relevant? Well, flow aside; the veins have the greatest TOTAL amount of blood of any part of the circulatory system. Veins have approximately 2/3 of the circulating volume. This means that veins are the blood reservoir of the body, and as a reservoir, they can be called upon at any time to "fill" the rest of the system. Arteries hold the remaining 1/3 of the blood volume. Remember the bathtub example from the beginning of the chapter "same flow, different volumes"!

So, let's switch gears a little. In order to understand how flow is taking place, we need to understand how much fluid pressure is in our pipes if **NO** flow is taking place. This is what we call mean systemic filling pressure (MSFP). If we were to stop somebody's heart, and with NO flow taking place, we were to analyze the pressure within the blood vessels, this would be the MSFP. This number, presented in mmHg, is a direct representation of the TOTAL amount of volume in the circulatory system. The more fluid in the circulation, the higher the MSFP will be! The MSFP is represented by point 1 on line E in Figure 3.6B. MSFP will be measured in the RA and if there is no flow, that is, no venous return (y-axis), then it will tell us how much volume is being occupied by the fluid in the CV system. Now, in order to make the venous function curve, we need to get flow started. Imagine that we start up the heart slowly and flow begins to take place all along the circulatory system. As the heart starts pumping blood forward RA→RV→Lungs→LA→LV, the pressure within the RA starts to decrease. Imagine you have a drink with a straw and you're just about getting ready to take a sip. At first there is no movement of water from the glass to your mouth. Why? Because there is no pressure gradient from the glass to your mouth! So what do you do? You generate suction in your mouth by generating a negative pressure, thereby decreasing the pressure in your mouth with respect to the glass. This generates a flow gradient from the glass and into your mouth! Well, the RA is just that! As the heart starts pumping the pressure in the RA decreases with respect to the pressure in the rest of the system, thus you generate a flow gradient from the veins to the RA! This is what we call venous return. This means that the faster the heart pumps, the lower the pressure in the RA and the more venous return there's going to be (this is represented by the arrow next to line E). This is true up to a point at which the gradient to the RA is so strong that the great veins in the thorax collapse upon themselves and flow is subsequently just maintained but can no longer increase. (This would be the equivalent of you generating too much suction when taking a sip and having the straw collapse on itself, or your cheeks collapse on themselves and you end up with a fish face.) This basic make up of the venous function curve can be applied to different volumes (Figure 3.6B, curves D and F). If there is a decrease in TOTAL VOLUME, then there will be leftward shift of the venous function curve (curve F) and the MSFP will be decreased (point 2). The curve has the same slope as the normal curve (curve E) but at much lower values for venous return. Conversely, an increase in TOTAL circulating volume would be translated in a rightward shift of the normal curve (curve D). This increase in volume generates an increase in MSFP (point 3).

Key

The venous function curve is going to represent TOTAL volume status.

The reader might have noticed the parallels between Figure 3.6A and B. In fact, from our previous discussion, we can say that the RA pressure and the LA pressure are similar if not equal because of the same amount of flow that is going through both of them. (In reality there are some variations between these pressures but for explanation purposes we will assume they are the same.) This means that the x-axis for both graphs is the same. Interestingly, we can also generate a parallel between venous return (L/min) and work. Venous return is the amount of blood that is returning to the heart for a given amount of heart function (remember that we said as the pump moves faster the more venous return there is going to be?), as the heart beats faster, the greater the venous return. Work is the amount effort that the heart exerts in order to pump THE BLOOD THAT IS GETTING TO IT. This means that work represents the amount of volume that the heart is moving for a set amount of venous return. In short, the y-axis for both graphs is the same. If both axes are the same, then we can superimpose one graph on the other one. Figure 3.7 is a graphical representation of a vascular curve superimposed on a cardiac function curve.

We had previously defined that venous return is dependent on cardiac function (decreased RA pressure) and that cardiac function is

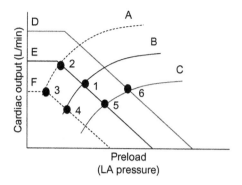

Figure 3.7 Cardiac output and its relation to venous return.

dependent on venous return (preload). As we superimpose the graphs, we're able to see this interdependence, the point at which the venous curve crosses the cardiac function curve represents our cardiac output! That means if we have a normal cardiac function curve, that is, our heart is working properly, and a normal venous function curve, that is, we have an adequate amount of circulating volume, our CO is going to be represented at point 1, and for the standardized patient, it will be approximately 5 L/min. Now comes the fun part! Let's play with the curves!

3.9 CHANGES IN CONTRACTILITY, VOLUME AND THEIR IMPACT ON CO

Starting off at point 1 (normal CO), let's say in the middle of reading this book you have an epiphany and realize that exercise is extremely important so you decide to jump on the treadmill and start working out. As soon as you start running, your heart will begin to pump faster because of increased sympathetic tone and stimulation. This means there's an increase in contractility! An increased contractility means moving from point 1 to point 2 in Figure 3.7. Additionally, we mentioned that an increased cardiac function was associated with greater venous return. If we look at point 2, we will be able to see that although we did NOT jump between venous function curves (i.e., our volume has remained constant because the MSFP has NOT changed), venous return has increased! This is a normal physiological response for exercise! Now imagine what would happen if circulating volume decreased as in hypovolemic shock. The first thing that happens is an acute decrease in volume. This acute decrease in volume means the venous function curve will shift to the left (curve F). If we remained with the same cardiac function as in the normal state (curve B), our cardiac output would drop precipitously (point 3). This is why the body, through the secretion of catecholamines among other hormones, increases cardiac contractility. This increase in contractility shifts our cardiac function from curve B to curve A. The point at which curve A and curve F intersect (point 4) is our new "steady-state" CO, which is much closer to normal than if there were no increase in contractility. (In reality, the changes that we outlined here happen simultaneously within a matter of minutes, but it is key to understand the exact progression of events.) Now, how does the body compensate decreases in contractility? Imagine our standardized patient has a heart attack and the LV wall is now hard and stiff. Contractility will obviously be decreased and if the venous function curve remains the same CO will drop (point 5). The body however does have a way of compensating. By increasing the TOTAL volume (shifting of the venous function curve to the right, from E to F), MSFP increases and PRELOAD increases. This increase in preload is able to increase CO at the expense of retaining huge amounts of volume (point 6).

Clinical Correlate

Patients with failure commonly suffer from edema. Edema is due to the accumulation of fluid in the EC space. The decrease in contractility leads to compensatory mechanisms leading to volume retention (shifting of the venous function curve and recovery of the CO) at the expense of retaining volume. All the extra fluid in the EC space will favor the formation of pulmonary edema if pressures between the different points of resistance are not kept in check. If pulmonary edema does ensue, the cause is volume overload, thus diuretics are the first-line therapy.

Now that we've understood how arteries, veins, and the heart regulated blood and fluid flow in the body, we will dwell further in the specific role that blood plays in the body regarding oxygen delivery!

3.10 CLINICAL VIGNETTES

A 70-year-old man, with a long-standing history of diabetes mellitus, smoking, hyperlipidemia, and obesity, comes to the ER after left-sided chest pain woke him from sleep. His initial work up showed an ST segment elevation in leads DII, DIII, and V4, V5, V6, and therefore the diagnosis of a left ventricular wall infarct was made. An echocardiogram shows limited motion of the left ventricular wall, with accompanying systolic dysfunction (an inability to pump blood effectively).

1. *In this patient, sympathetic tone would be:*
   A. Increased
   B. Decreased
   C. Remain unchaged

   Answer A. The clinical picture presented above of a left ventricular infarction is classically associated to a decreased cardiac output. Depending on the severity of the ischemic injury, CO can be affected only slightly or there can be a precipitous drop in CO leading to sudden death. As CO begins to drop, sympathetic tone is increased in an attempt to increase CO through increasing heart rate and contractility. The damaged heart however can only do so much. In turn, the veins and arteries will be subject to an increased sympathetic stimulation leading to both a left shift in the vascular function curve (owing to venoconstriction) and increased arterial blood pressure (owing to increased arteriolar constriction). The latter can be detrimental because it increases the work that the heart has to perform in order to pump blood. Initial treatment for an acute MI includes aspirin (to stop platelet aggregation and prevent further occlusion), morphine (to help with pain and decrease sympathetic stimulation from pain), oxygen (in an attempt to increase perfusion to the body), and nitrates (which at low doses promote venous dilation, therefore reducing preload and reducing the amount of work the infracted heart has to perform).

2. *After the administration of the nitroglycerin, the patient begins to feel better and the pain subsides. Taking into account the effect that nitroglycerin has on venous circulation, the mean systemic filling pressure in our patient will:*
   A. Increase
   B. Decrease
   C. Remain unchanged

   Answer B. Dilation of the venous system will increase the space where the blood needs to be distributed. In other words, the vascular space is larger! If 5 L were previously enough to fill our "vascular container" by dilating the veins, now we need 8 L to fill the vascular system. In this context, mean systemic filling pressure or MSFP will decrease. As the vascular system can now accommodate a lot more fluid owing the venodilation, the pressure of the fluid within the system will decrease. Therefore dilating the veins leads to a decrease in MSFP. Blood loss has a similar effect in that the container has more space than fluid, in the case of blood loss the container stays the same size but what changes is the amount of fluid within. Either way the alteration results in a large container with not enough fluid to maintain pressure within.

3. *The patient is taken to the operating room where a catheter is passed through the left coronary artery and the obstruction is removed. However, the patient suddenly takes a turn for the worse, becoming disoriented and subsequently unconscious. He is taken to the Intensive Care Unit, where his blood pressure and cardiac output are monitored. The monitor shows progressive increases in the LA pressure over a period of 1 h, but there is no significant change in cardiac output. Why would there be no change in the cardiac output in spite of increased LA pressure?*

   A. There isn't enough blood in the system to keep cardiac output elevated.
   B. The administration of nitroglycerin dropped venous return abruptly and thus cardiac output.
   C. In this case, LA pressure is a poor indicator of cardiomyocyte length, and therefore changes in LA cannot be equated to changes in cardiac output.

   Answer C. The Frank–Starling mechanism of the heart states that increased distention of the cardiomyocyte will lead to a more forceful contraction. This is why increased venous return is equated with increased cardiac output, as increased venous return will distend a healthy heart prompting a more forceful contraction and an increase in cardiac output. In the clinical setting, it's impossible to measure cardiomyocyte length, and thus we use pressure as an indirect indicator of cardiomyocyte length. In this case however, the patient is suffering from left ventricular dysfunction secondary to ischemia. The loss of ventricular function means that an increased arrival of blood the ventricle will not lead to stretching of the cardiomyocyte. A ventricle which cannot stretch, will have a decreased compliance and will have increased pressure and will lead to increased pressure in the LA without a corresponding increase in cardiac output.

CHAPTER 4

How Fluid Aides Oxygen Distribution: Role for Blood

4.1 THE NEED FOR FLUID MOTION IN THE BODY: ROLE OF BLOOD

As organisms developed from unicellular to multicellular and onward, they changed the way they produced energy. They began to "realize" that they could generate a lot more energy with the help of oxygen than without it (look up information on *aerobic metabolism* if you're curious to learn exactly how). Unicellular organisms previously had an advantage in that their entire cell membrane surface area was in contact with earth's atmosphere and could receive adequate amounts of oxygen by simple diffusion. Similarly, they could excrete wastes by either diffusion or dumping waste outside of the cell membrane. Multicellular organisms, however, have an increased number of cells with a decreased surface area that's in contact with the environment. This limits the capability of each cell to obtain oxygen from and dump carbon dioxide. Think about it this way, in the human body, if cells were required to obtain their oxygen directly from the environment, only the outermost layer of the skin would be oxygenated. All the internal organs would die because they have no access to air. Nature's way of solving this problem was developing a fluid medium to carry oxygen around to all the tissues. We've talked already at some length about ways fluid moves around. Now we're going to talk briefly about what that fluid is, and what are the most basic and important things to know about it. That fluid is blood.

4.2 BLOOD

First, a warning: blood is an *incredibly* complex topic. There are a multitude of journals and books dedicated to its study. There is an entire subspecialty of medicine that deals with nothing but blood. A recent study estimated that there are some 4,000 compounds within the human blood. These compounds include gases, water, electrolytes, fats, carbohydrates, proteins, antibodies, infection-fighting cells, and hormones. Clinically, we commonly only test for somewhere around 20 such compounds. However, we are going to choose just one aspect of blood to focus on, and that's delivery of oxygen through blood flow. This is because we feel it is the single most important role of blood, and one that is of paramount importance clinically. A human being can survive for weeks without food and days without water. How long could he survive without oxygen, though?

With that aside, let us begin our discussion of blood. Blood consists of two major components: (1) plasma and (2) cells (mainly red blood cells also called erythrocytes or RBCs). Plasma is what we've been referring to when we mention intravascular volume. Think of plasma as the fluid that the cells are bathed in. It is the fluid that RBCs float in when traveling through veins and arteries. On average, the male human body contains around 70 ml/kg blood. This corresponds to almost 5 liters of blood (visualize this amount in your mind's eye: all of the blood in your body would only fill up 2½ × 2 l bottles of soda!); 55% of this blood is made of plasma and about 45% are cells. These 5 liters of blood circulate one full turn within your body in just under 1 min at rest, and with intense running, blood can circulate one full turn in as little as 20 seconds!

Now, if blood were a simple fluid without cells, it would only be able to carry a small amount of oxygen. Why? Well, you should already know an answer to this question. Diffusion! We learned about it back in the first chapter. Remember that the air in our earth's atmosphere is only about 21% oxygen and 78% nitrogen. So the body needed to come up with a way to efficiently be able to "pick up" the oxygen in the lungs, then turn around and "drop it off" in the peripheral tissues. The body's rather elegant solution was creating a protein called hemoglobin. Thankfully, hemoglobin excels at reversibly binding oxygen in the lungs and releasing the oxygen in the periphery.

Hemoglobin (Hb) is 99% of the dry weight of a single RBC in total and there's approximately 1.5 lb of hemoglobin in the body! It is a something called a "metalloprotein." Sort of like a cyborg protein. It's part metal, part protein. The metal that's incorporated into Hb is iron. It is this metalloprotein that gives blood its red color when oxygenated (or purplish/blue when not oxygenated). There are four binding sites per hemoglobin molecule that can grab onto oxygen. A single gram of hemoglobin can bind 1.34 mL of oxygen. This is a 70-fold increase in oxygen-carrying capacity compared to oxygen dissolved in blood plasma alone! Now, if 1 g of Hb can bind to 1.34 mL of oxygen and if we were to saturate all the hemoglobin in the body with oxygen, then what would this amount to?

There are approximately 5 L of blood in the human body, and there are 15 g of hemoglobin per deciliter or per 100 mL. If there are 10 dL in 1 l (10×100 mL = 1 L; 50×100 mL = 5 L), we end up with something like this:

$$1.34 \text{ mL } O_2 \times 15 \, g/dL \times 50 = 1,005 \text{ mL of oxygen}$$

This means that if all the hemoglobin in the blood were fully saturated with oxygen, it would have approximately 1 L of oxygen!!!

Clinical Correlate

Why do we use the units grams/dL? Well, it's as arbitrary a unit as any other. But when you are working in the hospital, you will notice that one of the most common tests is called a complete blood count (or CBC). One of the four components of a CBC is hemoglobin, and the units are displayed in g/dL! This an actual measure of hemoglobin. The second component you'll notice related to amount of RBCs is a something called hematocrit. Hematocrit is basically just the percentage of RBCs vs plasma. You can spin the whole blood down in a centrifuge and the heavier cells separate and make their way to the bottom of the test tube. One can then look at what percent is red and what percent is not. As we said above, hematocrit is usually 45%. You'll notice that normally the ratio between hemoglobin and hematocrit is 3:1; 45% RBCs by hematocrit: 15 g/dL of Hb. That's because even though hemoglobin makes up an astounding 99% of an RBC by dry weight—in part because RBCs are one of the only cells in the human body to not contain a nucleus—it makes up 30% of the wet weight. If you round that up to 33%, then you see for each RBC, Hb is 1/3 of the "volume" of an RBC, making hemoglobin $\sim 1/3$ the total hematocrit. This can change if there is more or less fluid within the cell.

It may be helpful during our blood discussion to think of the vasculature as a railroad track and the RBCs as train cars on a commuter train. You can think of oxygen as a day shift workers heading into work and carbon dioxide as night shift workers heading home for the day. Both board the train. Some have a seat in chairs, and some prefer to stand. Oxygen gets on at the lung station and takes the train into work. It does work with cells out in the periphery, and it later makes its way home on another train, tired and in the form of carbon dioxide.

The engine for the train is the heart, which was discussed in the last chapter. Cardiac output determines just how quickly the train arrives at its destination. You can think of the chairs on the train as being hemoglobin chairs. As we said, hemoglobin is an oxygen-carrying protein that is highly expressed in RBCs. Hemoglobin has the ability to reversibly bind to and dissociate from O_2, which renders it an efficient way for oxygen to stay seated within the RBC, but also to get up and leave when it arrives at its stop. Hemoglobin requires the presence of iron (think: metal chair) in order to bind oxygen. Exactly how the iron binds oxygen is beyond the scope of this book. But a quick overview is that there are four binding sites on each hemoglobin molecule for oxygen. Imagine these binding sites like four seats in all bunched together in a row. As oxygen loads onto the chairs, it has to be seated in a certain configuration, starting with the window seat then lastly seating the aisle seat. Thankfully, oxygen is polite and fits neatly into one of these chairs. When all four oxygen molecules are seated, they all get along swimmingly. In other words, hemoglobin is somewhat more stable when the configuration has all four units bound to oxygen. However, should a person in the middle need to get up (i.e., a single oxygen molecule off-load), then all the other oxygen molecules have to get up to let that person out. In other words, once it releases one oxygen, it will easily off-load additional amounts of oxygen as the configuration becomes less stable with oxygen level leaving. The key concepts to know about hemoglobin are that it seeks out oxygen and holds on to it, but not so aggressively that it doesn't want to let it go. It's happy to be used for only a small amount of time, and then to allow oxygen to leave when the cell "realizes" it's approaching its stop.

Some of the signals that the RBC is approaching its stop are signals that will modify the binding of O_2 to Hb including oxygen concentration, blood pH, temperature, concentration of CO_2, and a metabolite from the glycolytic pathway in the RBC: 2–3 DPG. You'll learn more about these individual factors when you learn pulmonary physiology, but all you really need to know is that all of these above factors are expressed in active tissues! Isn't that convenient! These cells show signs they're working hard and hemoglobin can biochemically gift them oxygen without even asking. So in addition to all the flow mechanics we mentioned in previous chapters, more oxygen can be delivered to tissues in need simply by hemoglobin recognizing their metabolic need.

This relationship between amount of oxygen (in mmHg) and saturation of Hb (how much oxygen Hb is carrying) is plotted in graph form in Figure 4.1. Here, we can see that at high O_2 pressure, saturation will be equally high. As the O_2 pressure decreases to approximately 40 mmHg, the saturation only drops to 75%. This means that a 60 mmHg drop in O_2 only reduces Hb O_2 saturation by 25%! The body's normal functions occur between these two points. The pressure of oxygen in arterial blood is approximately 100 mmHg and the pressure of oxygen in venous blood is around 40 mmHg. This variation in oxygen content will only slightly affect the saturation. What does this represent in terms of oxygen transport? Let's take a look at our formulas:

$$\text{Oxygen delivery in ml/min} = (\text{CO}) \times (\text{Hb } g/\text{dL} \times 1.34 \text{ mL}) \times \% \text{ Sat Hb } O_2$$

In the arterial blood:

$$50 \text{ dL} \times 1.34 \text{ mL}/O_2 \times 15 \text{ } g/\text{dL} \times 100\% = 1,000 \text{ mL of oxygen/min}$$

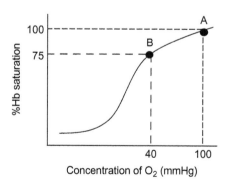

Figure 4.1 The oxygen–hemoglobin saturation curve. We generally "live" between points A and B, where A represents arterial saturation of oxygen and B the venous saturation, and the difference between both points represents oxygen extraction by tissues!

(Remember normal cardiac output is approximately 5 L/min and there are 50 dL in 5 L.)

This represents the amount of oxygen that gets to the tissues, but how much is actually being extracted? The easiest way is to use the same formula to figure out the oxygen content of venous blood and then subtract the result from the arterial blood. Let us rephrase, if you know how much you start with (arterial content of O_2) and know how much you end up with (venous content of O_2), you'll know how much you consumed.

So the oxygen content of venous blood:

$$50\ dL \times 1.34\ mL/O_2 \times 15\ g/dL \times 75\% = 750\ mL\ of\ oxygen/min$$

(Remember the saturation is approximately 75%.)

Thus:

$$Arterial\ O_2\ content - Venous\ O_2\ content = Oxygen\ extraction$$

$$1,000\ mL - 750\ mL = 250\ mL/min$$

This means that for every minute of the 1,000 mL of O_2 carried by the blood, the body consumes approximately 250 mL.

So, if we analyze our formula and train analogy a little closer, we can see that there are **only three things** that the body can do to increase oxygen delivery to its organs:

1. **Increase cardiac output**—If you take a bullet train to your destination instead of the slow local train, you can imagine you'll move more people to and from their destination! (Increase blood flow means more blood cells get loaded up with oxygen and more of them make it to their destination over set period of time.)
2. **Increase the concentration of Hb**—In the body, this happens slowly over time and is one of the main reasons you need to consume iron. Or you could go the Lance Armstrong route and increase your Hb concentration by "blood doping"—typically either giving yourself a hormone that regulates RBC production, erythropoietin (EPO), which will be discussed in next chapter, or by simply transfusing yourself with blood.
3. **Increase the percentage saturation of Hb with O_2**—This means increasing the amount of oxygen that the Hb is carrying.

This is it! There is nothing more that we can do! This is why it becomes increasingly important to thoroughly understand what each of these variables entails, because when we're standing at the patient's bedside, all we can modify are these three variables!

We have already seen several ways of increasing CO. Remember from the previous chapter that if you increase venous return, contractility or heart rate you can increase cardiac output, and all of these events are secondary to catecholamine secretion. Additionally, we can add fluid to the body and this will in turn (depending on the Na^+ and water balance) be distributed throughout the body, including the EC space. We will analyze the exact response to decreased CO in a future chapter but for now, let's just focus on the fact that if we increase CO we increase oxygen delivery. **Let's say it again: increasing CO increases Oxygen delivery!! This is one of the MAJOR POINTS OF THIS WHOLE BOOK!!**

The second and third components of the formula are a lot more straightforward. If we increase the amount of Hb in the body, we increase the oxygen-carrying capacity. This is why we give people blood transfusions. By transfusing packed red blood cells (PRBCs), we increase the total content of hemoglobin and thus the number of transport proteins available to shuttle oxygen around. (Blood is divided into components to allow for a more efficient use. Whole blood has a relatively short expiration date, while blood divided into red blood cells (PRBCs), plasma, and platelets last a lot longer.)

Clinical Correlate

Hemoglobin and hematocrit are the two components of a CBC that are used to assess oxygen-carrying capacity. Regrettably there is no straightforward answer to the question: When do we transfuse someone? We must take all variables into account including patient age, initial cause of blood loss (acute vs chronic), heart function, and overall fluid status before we decide to transfuse. As a general rule, in the absence of other modifying variables, hemoglobin of less than 7 g/dl is a pretty strong indicator that a patient needs blood.

Now that we have those extra transport proteins, what can we do to make them work adequately? We make sure that they are completely saturated with oxygen so that not one free binding site for Hb goes to waste.

This is achieved by increasing the oxygen saturation of Hb, by increasing the amount of oxygen in the air that our patients are breathing.

Clinical Correlate

FiO_2 is the clinical term with which we refer to fraction of inspired oxygen, that is, how much of the air we breathe is actually oxygen. Atmospheric oxygen levels are 21%, this means that out of every 100 mmHg of atmospheric pressure, 21 mmHg are due to oxygen. We can increase the FiO_2 in the clinical setting through the use of many devices including nasal cannulas, nonrebreather masks, and even endotracheal tubes. We can modify the FiO_2 in order to increase the % of Hb saturated with oxygen. By increasing the FiO_2, we attempt to ensure that all the O_2 binding sites on Hb are being utilized.

In summary, we have seen that blood plays a key role in transporting oxygen from the lungs to the tissues, which is represented in the formula:

$$\text{Oxygen delivery in mL/min} = (CO) \times (\text{Hb } g/dL \times 1.34 \text{ mL}) \times \% \text{ Sat Hb O}_2$$

Thus, we can alter only three variables if we want to increase oxygen delivery, cardiac output, hemoglobin quantity, and saturation. Of these factors, we've consistently hammered CO as key in this whole process. As you now know, CO is a product of EC fluid distribution, heart function, and as we will see in the next chapter, regulation by the kidneys!

4.3 CLINICAL VIGNETTES

A 55-year-old male comes to the physician for a schedule physical and laboratory tests. The results from his CBC show that he has a hemoglobin of 10 g/dl (normal 13−15 g/dl).

1. *The decreased hemoglobin concentration in this patient has a direct effect on oxygen delivery. If the patient is not placed on oxygen, what other compensatory mechanism could be found in this patient that would attempt to offset the decreased oxygen transport by the low hemoglobin level?*
   A. Increase in the saturation of hemoglobin
   B. Increase in blood volume

C. Increase in cardiac output.

Answer C. An increase in cardiac output, likely through tachycardia (heart rate > 100 beats/min) would be an efficient way to increase oxygen delivery. Remember that the oxygen delivery formula is

$$\text{Oxygen delivery} = CO \times Hgb \, (g/dL) \times \text{Saturation}$$

In this case, the saturation of hemoglobin would not change, because, at room air (21% oxygen), hemoglobin is likely saturated at 99%. Alternatively, since we are not giving the patient any supplemental oxygen, the saturation of hemoglobin will not increase. Increases in blood volume would necessarily increase hemoglobin concentration. This compensatory mechanism can be seen in certain patients with respiratory problems that lead to increased blood volume. The most likely compensatory mechanism in this patient, tachycardia, would increase cardiac output (remember CO = heart rate × stroke volume).

2. *As expected, the patient's heart rate is 117 at rest (normal heart rate is 60–100). An effective way to decrease this patient's heart rate and still maintain oxygen delivery would be:*
   A. Administer beta blocking medications
   B. Transfuse the patient with blood
   C. Give epinephrine.

Answer B. The patient's heart rate is elevated secondary to poor oxygen delivery to the tissues. Given the decreased hemoglobin concentration, the heart rate increased as a means to increase cardiac output and therefore compensate the decreased oxygen delivery. Administering beta blocking medications would decrease the heart rate but would not maintain oxygen delivery. By solely decreasing heart rate, cardiac output would be decreased and oxygen delivery would equally be reduced. Giving epinephrine would increase oxygen delivery but would also increase heart rate. The only answer that would increase oxygen delivery, while decreasing heart rate, is giving a blood transfusion. By giving blood, the concentration of hemoglobin will increase, thereby increasing oxygen delivery. Once oxygen delivery increases through the increase in hemoglobin, heart rate will decrease as compensation.

3. *The patient then asks the physician if having a tank of supplemental oxygen at home would help with his problem. The most appropriate reponse is:*
   A. Yes, having supplemental oxygen will help because it will significantly increase the amount of oxygen being transport by the hemoglobin.

B. No, supplemental oxygen will not help because the hemoglobin is already carrying as much oxygen as it can.

Answer B. Hemoglobin is normally saturated to 99–100% at a normal atmospheric pressure of oxygen, which corresponds to approximately 21% of the air in the atmosphere. An increase in the fraction of inspired oxygen (FiO_2) will only increase the saturation of oxygen in hemoglobin that has a low saturation, that is, hemoglobin that is not carrying all the oxygen it can! Hemoglobin represents the train and oxygen represents the people. You can try to load as many people as you can in the train, but if all the seats are taken, it doesn't matter how many people are waiting, the train is already at maximum capacity! Similarly if hemoglobin is already saturated at 100%, there is no room for more oxygen molecules to bind. Increasing the FiO_2 will only work in patients in which hemoglobin is not carrying all the oxygen it can, that is, there are free seats on the train!

CHAPTER 5

Fluid Handling by the Kidneys

As we discussed previously, changes in fluid volume and movement have a huge impact in cell size, cardiac output, and even oxygen delivery! Changes in circulating volume and total body water can be acutely compensated in a number of ways including regulating resistance (vasoconstriction, vasodilation), flow (cardiac function), or generating intra−extracellular fluid shifts. These changes, however, will provide a temporary solution to a much bigger problem: what is causing these fluid shifts, is there too much volume, not enough? The brain can tell the body is thirsty by recognizing changes in body fluids, it can help you find water and physically control the process of drinking. Your small intestine can help you absorb water into the bloodstream. But none of this would matter without the kidney. The kidney is the only organ that can regulate the TOTAL AMOUNT of water and electrolytes in the body. The basic mechanisms behind how the kidney manages to do this will be summarized and explained; however, as with other chapters, remember to refer to a physiology textbook for more in-depth answers. Our goal with this chapter, as with all the chapters in this book, is to give you a springboard to understanding. We will cover the kidney in some depth, but know that it is often true in medicine that better understanding of the basics and how they're integrated will make the more advanced stuff a lot easier to understand and remember.

The kidney is the body's filter. It filters blood to produce urine, which contains the "waste products" found in blood. It achieves this through three basic functions: filtration, reabsorption, and secretion. Arterial blood will arrive at the kidney through the renal arteries and a series of high-pressure conduits that will aid in filtration. After the blood is filtered (which is a massive and basically unregulated process), a mode of "purification" takes place, that is, reabsorption and secretion. These actions are highly regulated and specific, in that even the most minute ionic concentrations can be altered.

How is this achieved? Imagine that I put a bowl of color beads in front of you. Inside the bowl, there are approximately 5,000 beads of all different sorts of colors. Now I give you these instructions:

- At the end of the day, I want 3,000 red beads, 1,000 blue beads, 500 pink beads, and the remaining 500 purple beads go into the trash.

This seems rather difficult to do. ... But what if I change the instructions a little:

- From the bowl of beads in front of you, I will give you a spoon that extracts exactly 100 beads and from each spoonful (filtration), I want you to keep all the red, blue, and pink (reabsorption), and throw all the purple beads in the trash (secretion), one spoonful at a time.

This seems a lot more manageable. (The end result is the same!) If you deal with small fractions of the whole, it's a lot easier to pick and choose what you want and what you don't want. This is what the kidney does! By constantly filtering small amounts of blood and deciding what to do with that small amount of blood, it regulates the entire system. Not only this, but by filtering these small amounts of blood, it will be able to circulate the same volume a great number of times, thus increasing the specificity. (We have approximately 5 L of circulating blood volume; the kidney however filters close to 180 L per day or **125 mL/min**. This means that the entire blood volume is filtered 36 times throughout the day.)

Key

The standard glomerular filtration rate (GFR) is 125 mL/min.

How does the kidney achieve this? The functional unit of the kidney is the nephron. A nephron is basically a group of cells that decide to form a tube. The difference between each cell type and the specific anatomical distribution of each cell type is essential to nephron function. There are approximately 1 million nephrons per kidney, and this number begins to decline (in normal individuals) after around 40 years of age at a rate of approximately 1% per year. Of these, 1 million nephrons per kidney around 80% are located in the renal cortex—the outermost portion of the kidney. These are called cortical nephrons. The remaining 20% span the whole kidney from cortex to medulla—the inner portions of the kidney. These are called juxtamedullary nephrons (*juxta* means next to, *medullary* medulla). **The juxtamedullary nephrons are the ones in which the concentration and dilution of urine primarily take place.**

Figure 5.1 is a schematic representation of a nephron. Each nephron segment that will be discussed is labeled with a letter. From proximal to distal they are:

A—glomerulus
B—proximal convoluted tubule (PCT)
C—descending loop of Henle (DLH)
D—thick ascending loop of Henle (TALH).
E—distal convoluted tubule (DCT) and connecting/collecting ducts (CNT/CD)

Additionally, the dotted arrow points to a very specific region of the nephron called the macula densa (MD) which will be discussed later on in the text. However, the reader has to notice that the MD is located at a very specific portion of the nephron in which it allows the communication between the proximal and distal portions of the nephron.

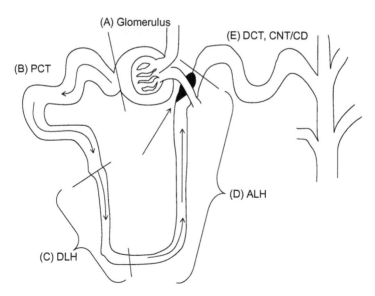

Figure 5.1 The nephron—the functional unit of the kidney is subdivided into functional segments: (A) glomerulus, (B) PCT, (C) DLH, (D) ALH, (E) DCT, CNT/CD. Additionally, a specific portion of the TALH (arrow) is known as the MD.

5.1 THE GLOMERULUS

The first portion of the nephron is composed of several structures. The glomerulus represents the filter of the kidney. Starling's forces (discussed in Chapter 2) are going to play a key role in determining filtration, and the entire glomerulus is built around how these forces are going to interact in order to produce a filtrate. Let us explain. Imagine a strainer. When we're making orange juice and we want no pulp in our orange juice, we use a strainer. How does the strainer work? Well we pour in the whole goop that we just got from squeezing our oranges and the strainer will hold on to the pulp, but will allow the juice to continue through to the glass. The juice is not only water: it's also sugar, natural colorants, ions, vitamins, etc. So the strainer is just really filtering the massive chunks of pulp that are too big to go through. In a sense this is exactly how the glomerulus works. It will strain the "juice" (blood plasma, the 55% we were talking about earlier) and leave the "pulp" (RBCs, WBCs, and proteins) behind. Thus the only thing which is freely filterable is the plasma and anything dissolved in it (remember what we call this? Interstitial fluid!!).

Let's study the makeup of the glomerulus a little more closely to see how its form explains its function. Figure 5.2A shows us the glomerulus with its major components: the Afferent arteriole (where blood Arrives), the glomerular capillary tuft, the Efferent arteriole (where blood Exits to provide O_2 to the rest of the nephron and consequently the entire kidney), and Bowman's capsule (which receives the filtrate and channels it to the PCT). The whole thing, when sliced in half, looks like a roundabout connecting two roads, the afferent and efferent arteriolar roads. The blood is arriving to the glomerulus by the afferent arteriole, and whatever is not filtered is picked up and exits via the efferent arteriole.

Key

The remaining blood that bypasses the filter and leaves via the efferent arteriole is the blood that will provide O_2 to the kidney cells! This means that if less blood goes through the efferent arteriole, the O_2 delivery to the kidney parenchyma decreases!!

The glomerular capillaries are a high-pressure capillary bed that is formed between the afferent and efferent arterioles. This is the strainer!

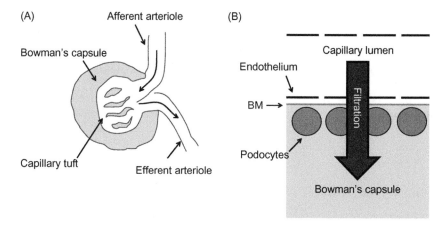

Figure 5.2 The glomerulus (A) is formed by the capillaries which arise from the afferent arteriole and then coalesce to form the efferent arteriole. They are known as the glomerular capillaries. Blood arriving from the renal artery enters the capillaries through the afferent arteriole and blood moving toward the peritubular capillaries exits the glomerulus through the efferent arteriole. The filtration barrier (B) is made up of three components: the fenestrated endothelium (it has got holes in it), the glomerular capillary basement membrane (BM) which has a net negative charge to prevent filtration of proteins, and the podocytes which are cells that surround the glomerular capillaries (they too have a net negative charge).

Whatever the capillaries filter will be caught by Bowman's capsule and channeled through the tubules! (Bowman's capsule is the equivalent of the glass in our OJ example.) Now, let's take a closer look at the glomerular capillary membrane.

Figure 5.2B is a schematic of all the layers that a molecule must traverse through to get through the capillary wall. From glomerular capillary lumen to Bowman's capsule, a molecule must go through: (1) capillary endothelium, (2) capillary basement membrane, and (3) podocytes. Now, for all intents and purposes a strainer has holes, so the glomerular capillaries must have holes as well to allow for filtration. These holes are in the endothelium and basement membranes (they are generally referred to as fenestrations). Now additionally, the basement membrane has an invisible electric barrier as well, as it's composed of a high concentration of negatively-charged molecules and proteins. The combination of the fenestrations and the net negative charge of the glomerular basement membrane make up the filtration barrier, thus filtering molecules and blood components in two ways:

1. Anything bigger than the fenestrations will not be able to go through, for example, cells (RBCs, WBCs), and some proteins.

2. Anything that has a negative charge will be repelled by the net negative charge of the basement membrane and will not be filtered. Note that many blood proteins carry a net negative charge, so this is a double barrier against the smaller charged proteins.

This sets up a filtration system that, under normal circumstances, will not filter:

1. Cells
2. Proteins.

But will freely filter:

1. H_2O
2. Electrolytes
3. Glucose
4. Small, noncharged molecules, for example, creatinine and urea.

So, what actually ends up in Bowman's capsule after filtration has taken place, is water with electrolytes, glucose, and small molecules in a concentration that is very similar to the electrolyte concentrations found in the interstitial fluid! How exactly does the glomerulus achieve this? Well, in large part you already know: Starling's forces!

Figure 5.3 is a schematic representation of Starling's forces and their relationship in the glomerular capillary. In order for the glomerulus to filter appropriately, there needs to be a high hydrostatic pressure inside the glomerular capillaries. This is achieved as the branches of the renal artery give off a series of high-pressure conduits that will favor an increased hydrostatic pressure in the glomerular capillaries, two of which are the afferent arteriole and the efferent arteriole. In fact, the glomerular capillary hydrostatic pressure (GCH) is close to 70 mmHg! This GCH will favor the leakage of fluid from inside the capillaries toward Bowman's capsule (out of the strainer and into the glass). On the contrary, Bowman's capsule hydrostatic pressure (BCH) will oppose the exit of fluid from the capillaries.

In addition to hydrostatic pressure, the pressure of proteins (oncotic pressure) also determines fluid leakage or retention from the capillaries. However, as stated earlier in the chapter, there are hardly any proteins in the ultrafiltrate because of the net negative charge in the glomerular capillary basement membrane. This means that proteins

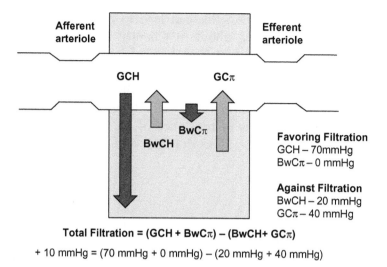

Total Filtration = (GCH + BwCπ) − (BwCH+ GCπ)

+ 10 mmHg = (70 mmHg + 0 mmHg) − (20 mmHg + 40 mmHg)

Figure 5.3 Filtration is the product of the interaction of Starling's forces within the glomerular capillary and Bowman's capsule. The forces favoring filtration are the GCH and Bowman's capsule oncotic (BwC$_π$). The forces opposing filtration are BCH and the glomerular capillary oncotic (GC$_π$).

will be concentrated inside the glomerular capillaries. This increased oncotic pressure within the capillaries themselves will oppose the fluid leakage, that is, proteins in the capillaries oppose filtration!

Clinical Correlate

In a clinical condition known as nephrotic syndrome, you have for one reason or another a more porous sieve. The glomeruli allow more proteins such as albumin to pass into the ultrafiltrate and eventually get urinated out of the body. This results in a decrease in plasma oncotic pressure. This decrease in plasma oncotic pressure not only causes more fluid to be filtered out of the capillaries at the glomeruli, but throughout all the capillaries in the body! This increase in capillary filtration ends up causing edema (swelling) throughout the body despite an increase in urine output! We will discuss other problems in filtration toward the end of the chapter after we learn more about the rest of the kidney.

Now then, we said that proteins within the glomerular capillaries cause an increase in oncotic pressure and will oppose filtration. However proteins are not something that we can quickly modify in order to alter filtration.

Something that can be modified, however, is the hydrostatic pressure within the glomeruli (GCH). By changing the resistance of the afferent or efferent arterioles, this can lead to changes in the GCH. Remember what we talked about in Chapter 2 regarding pressure, resistance, and flow? We explained how changes in resistance can rapidly affect pressure and flow upstream and downstream. Your body applies this principle everyday when it wants to decrease or increase the amount of plasma that gets filtered! Figure 5.4 is an illustration of how changes in resistance of the afferent arteriole (B) or efferent arteriole (C) affect GCH, and subsequently affect filtration.

Key

If we increase the resistance of an arteriole, capillary hydrostatic pressure increases upstream of that point of resistance; if we decrease the resistance the hydrostatic pressure will increase downstream of that point of resistance.

If we take Figure 5.4A to be steady-state conditions and we decrease afferent arteriolar resistance (Figure 5.4B), there will be an increase in the GCH and thus increased filtration. Likewise, if we increase resistance at the efferent arteriole (Figure 5.4C), this will increase the GCH and thus increase filtration. Regulation of filtration

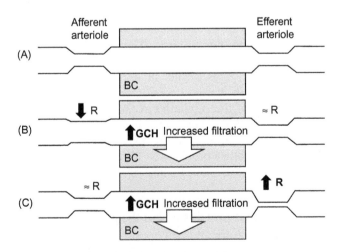

Figure 5.4 Changes in either afferent or efferent arteriolar resistance have an impact on filtration. A decreased afferent resistance (B) increases flow to the glomerular capillaries (GC), thus if efferent arteriolar resistance remains the same, the GCH increases, favoring filtration. If afferent arterioral resistance doesn't change but efferent arteriolar resistance increases (C), this decreases flow to the peritubular capillaries and increases GCH (remember from retrograde flow from Chapter 2), thus favoring filtration.

can therefore be extremely varied, as there are a number of hormones and vasoactive substances that can modify arteriolar resistance. The simplest—albeit not the most comprehensive—way to understand regulation of filtration is with the concept of local vs systemic regulation of filtration, which we will explain in depth further on in the chapter. However, we will advance a couple of key notions. Local regulation involves each nephron regulating its own flow! (i.e., adjusting GFR one nephron at a time). Systemic regulation involves a signaling system that will supersede each nephron's "free will," thus regulating global renal filtration in order to accommodate the needs of the entire body (i.e., adjusting renal blood flow and total renal GFR as a consequence).

5.2 TUBULAR FUNCTION

After filtration has occurred and the filtrated plasma has pushed its way into Bowman's capsule, it will continue to advance through the tubules, so that the highly regulated and specific processes of reabsorption and secretion can take place. In order for the tubules to function properly, we must address certain elements that are common throughout the different tubules present in the nephron. One of the most important elements in tubular physiology, as in all cells really, is the Na^+/K^+ ATPase. Proper functioning of this ATPase can be seen in Figure 5.5. The ion pump in this example is located in the PCT. Note that the Na^+/K^+ ATPase pump in this figure is located at the basolateral membrane. This is the outer part of the membrane of one of the cells making up the wall of the nephron's PCT. This part of the cell membrane faces outward toward the interstitium and rest of the kidney parenchyma. Note that this specific interstitial space may also be called the basolateral space. Now, the efferent arteriole gives off a series of low-pressure conduits (the peritubular capillaries) that accompany the tubules and provide the O_2 necessary for Na^+/K^+ ATPase function and consequently adequate tubular functioning. The basolateral space contains the peritubular capillaries! So think of the basolateral membrane as the membrane that faces the blood that just flowed past the glomerulus! This is true of all the tubular cells the entire length of the nephron! Throughout the entire kidney, the Na^+/K^+ ATPase will do what it always does: pump Na^+ out of cells and into the blood.

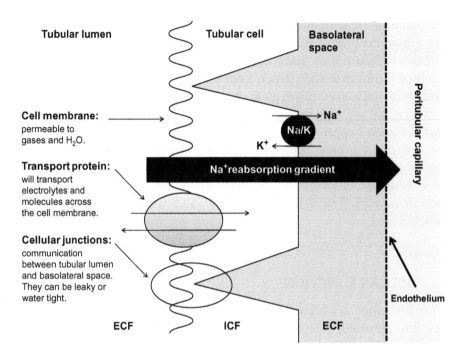

Figure 5.5 Tubular function.

Key

The basolateral membrane faces the peritubular capillaries. The peritubular capillaries are the capillaries downstream of the efferent arteriole of the glomerulus and run along the nephron's tubule. These capillaries are the vessels that feed the nephron with oxygen to allow the ATPase and thus the tubule to work!

In order to properly understand tubular function, we must understand the particulars of what tubular fluid is and where it's going. Tubular fluid is the plasma that was filtered by the glomerulus, and as we mentioned previously it's the "pulp-less" OJ, that is, water, with electrolytes, glucose, and small molecules. This fluid is now confined to the tubules, which are a part of the extra cellular fluid (ECF) and must find a way back into the blood. With this in mind, let's analyze a generic tubule with the basic components set in place (Figure 5.5). As you can see, the path from the tubular fluid back to the blood is lined with obstacles! After filtration, tubular fluid will move along the tubular lumen. The tubules themselves are lined by tubular cells, which express the Na^+/K^+ ATPase on the basolateral side. Tubular cells are joined together by either tight or leaky junctions which, depending on the segment of the nephron, will allow or inhibit the passage of fluid, electrolytes, and small molecules to the basolateral space. A large number of compounds, however, need to be shuttled directly through the tubular cell! Because the cell membrane is not readily permeable to electrolytes, the expression of transport proteins is KEY in shuttling nonpermeable compounds across the cell! However, as we saw previously, transport doesn't just happen, there needs to be a gradient, and the tubules use one of three mechanisms to transport electrolytes and compounds:

1. Na^+ coupled reabsorption gradient generated by the Na^+/K^+ ATPase (by far the most important). As the tubular concentration of Na^+ approximates that of ECF (140 mEq/L) and the intracellular concentration of Na^+ is very low, transport proteins take advantage of the Na^+ drive to move all sorts of things across.
2. Charge—Through the differential transport of anions and cations, some regions of the nephron will have either negative or positive luminal charges favoring the movement of electrically charged molecules!
3. Primary transport pumps—When a gradient is not present but something needs to get transported, pumps will consume ATP and move stuff around regardless of the concentration gradient!

Tubular function is thus defined as the process that will reabsorb components of the tubular fluid from inside the tubular lumen, through the tubular cells and back into blood, and secrete waste products into the tubular lumen from the tubular cell. This is a highly organized and regulated process that deals with billions of individual molecules. The only way to deal with billions upon billions of molecules that need sorting is through a specialized hierarchy of functions that will generate as much of a steady state as possible. The tubular system is designed in such a way (Figure 5.6). From proximal to distal, the nephron will first deal with bulk reabsorption (1) massive iso-osmolar reabsorption in the PCT (70% of filtered load), (2) reabsorption of water and some electrolytes in the loop of Henle (25% of the filtered load), and (3) fine-tuning of Na^+, K^+, Cl^-, and acid–base reabsorption and secretion in the DCT and CNT/CD (5–10% of the filtered load).

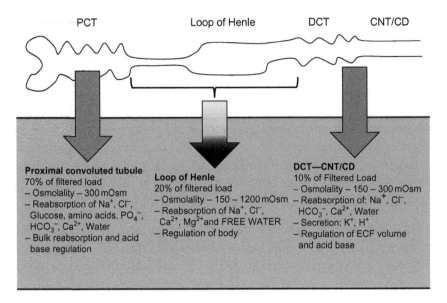

Figure 5.6 Tubular function along the nephron.

Clinical Correlate

Acute tubular necrosis (ATN) is the pathophysiological extreme of what happens when the tubular cells don't get enough O_2 from the peritubular capillaries. It's basically a tubular infarction (think "tubule attack" instead of "heart attack"). This happens when decreased blood flow to the kidney decreases perfusion of the tubular cells. ATN is a cause of acute kidney injury (also known as acute renal failure); in some cases the kidney can recover and regain some degree of normal function, however, the long-term consequences are still debated.

5.2.1 Proximal Convoluted Tubule

The concerted picture of reabsorption along the PCT is one of ISO-OSMOLAR reabsorption. This means that the same amount of solutes and water get reabsorbed. The osmolality of the fluid going in to the PCT (300 mOsm) is exactly the same as the osmolality of fluid leaving the PCT (300 mOsm). The only difference is the amount of water and solutes that are left! The PCT will reabsorb close to 70% of the filtered load, this means that out of the 125 mL/min that are filtered (remember GFR = 125 mL/min), approximately 87.5 mL will be reabsorbed in the PCT every minute. (If we put this in liters, the PCT will reabsorb 126 of the 180 L that are filtered by the kidney everyday). Put into simpler terms, the PCT's job is to reabsorb as much fluid and solutes as possible, leaving the rest of the nephron to deal with the problems of concentration, dilution, and volume regulation.

Expression of different types of apical membrane proteins involved in ion and other molecule transport is what defines the different cell types that make up the nephron's tubules. As we mentioned earlier, the first part of the tubules after Bowman's capsule is the PCT. Because the filtrate that arrives at the PCT comes from Bowman's capsule, this means that the concentration of ions and small molecules is almost exactly the same as that in the plasma (this translates into a concentration or approximately 140 mEq/L of Na^+).

PCT cells take advantage of the Na^+ gradient generated by the Na^+/K^+ ATPase to reabsorb the most certain ions and molecules first which, if not reabsorbed at the PCT, will end up in the urine because no other tubular cell is capable of reabsorbing them. As such, reabsorption of glucose and amino acids only takes place in the PCT, and it happens through cotransporters that move ions of Na^+, together with glucose, amino acids, or PO_4 (Figure 5.7). These transporters have a saturation point (don't worry, we won't go into transport kinetics). This means that they will become saturated if the concentration of glucose in the tubular lumen rises above 180 mg/dL.

Key

The tubular concentration of glucose is the same as that in plasma because glucose is freely filtered.

Clinical Correlate

Patients with Diabetes mellitus type 1 lack the insulin necessary for glucose to enter liver, fat, and muscle cells. This leads to an increased level of glucose in plasma. The increased level of glucose in plasma will result in increased glucose in the tubular fluid, leading to saturation of the Na−glucose cotransporter and subsequently lead to glucose turning up in the urine. As glucose has osmotic activity, glucose within the tubular fluid will "pull" water with it, and this will increase urine output. This provides an explanation for the "polyuria" and subsequent "polydipsia" that patients with DM1 can have.

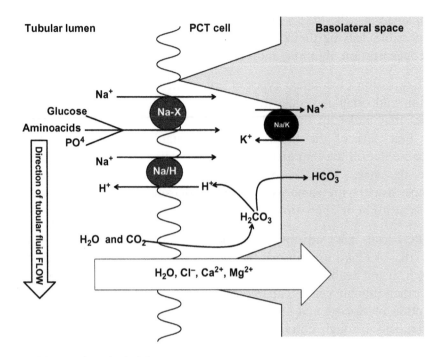

Figure 5.7 Proximal convoluted tubule.

The reabsorption of HCO_3^- is also major aspect of PCT function. In order to reabsorb HCO_3^-, you need to secrete H^+, and this is where it gets tricky (Figure 5.7). You will find other textbooks describing reabsorption of HCO_3^- and formation of new HCO_3^-. This has to do with whether the bicarbonate was initially filtered (i.e., it comes from the blood) or if it was formed inside the tubular cell and then transported to the blood. We will not dwell on this aspect and only mention total HCO_3^- balance. In the PCT, a great deal of the total HCO_3^- is reabsorbed, which means that any alteration in PCT function can lead to metabolic acidosis.

So far we have seen that the PCT is key in reabsorbing glucose, HCO_3^-, amino acids, and Na^+, through transcellular pathways. This, however, is not the end of the story. The junction between PCT cells is relatively weak, which means that paracellular transport can take place! Paracellular transport will be powered by water and a phenomenon called "solvent drag." As Na^+ is reabsorbed, the concentration of Na^+ on the basolateral side of the PCT increases, leading to an increased osmotic gradient for water from the tubular lumen to the basolateral side. Because there are leaky junctions between PCT cells, water will freely diffuse across in order to reach osmotic equilibrium. Water, however, has particles dissolved in it, like Cl^-, Ca^{2+}, and Mg^{2+}, which will move along with water and diffuse to the basolateral side. Imagine being in the middle of a crowd that starts to move, you will get pushed along no matter what you do. The same thing will happen to these ions as water begins to be reabsorbed. Additionally, all the proteins that were not filtered in the glomerular capillaries and were thus "concentrated" now manage to exert a pretty significant oncotic force to the PCT water, thus leading to increased reabsorption.

Clinical Correlate

There are two types of PCT diuretics: (1) osmotic diuretics, for example, mannitol and glucose and (2) carbonic anhydrase inhibitors, for example, acetazolamide. Osmotic diuretics work by osmosis, that is, pulling water into the tubular lumen using an osmotically active agent that can't be reabsorbed. Acetazolamide works by inhibiting the active transport of Na^+ associated with H^+ secretion. They are both weak diuretics.

5.2.2 Loop of Henle and Urinary Concentration

The osmolality of PCT fluid is the same as plasmatic osmolality, 300 mOsm. Through the process of urinary concentration, the body can reabsorb water, without necessarily reabsorbing solutes, thereby eliminating the waste products without increasing water losses. However, as there are NO primary water transporters (i.e., transporters that use ATP to move water), a gradient for water reabsorption must first be generated so that water retention will occur. The gradient for water reabsorption comes in the form of salt and urea being accumulated in the kidney's medulla by the juxtaglomerular nephrons. This means two things:

1. The reabsorption of water in the medulla of the kidney is driven by a Na^+- and urea-dependent high osmolality gradient between the medulla and tubules.
2. The reabsorption of water in this segment DOES NOT imply increased reabsorption of salt. In other words, the kidney generates a salt-dependent gradient for reabsorption of water, but reabsorbs A LOT MORE WATER than salt. So we can say that the basic function of the loop of Henle is to concentrate urine by favoring FREE WATER REABSORPTION.

Additionally, the loop of Henle plays a key role in regulating both the rate of glomerular filtration and blood pressure. Through sensing the arrival of NaCl, a specialized portion of the loop of Henle called the MD will give feedback to the glomerulus and regulate its flow! Now, first, we will focus on the water and salt reabsorption capacities and in order to conceptually understand how the kidney generates the gradient that allows for free water reabsorption, we must study the loop of Henle in reverse order.

(Before we begin, a brief note, our representation of the loop of Henle and peritubular capillaries is for schematic purposes only, the actual anatomic disposition of loops, capillaries and ducts is increasingly complicated and ultimately unknown. We will explain concentration and dilution by making an emphasis on the direction of flow, which is an accurate representation of the final effect of the loop of Henle.)

The loop of Henle is divided into two segments: descending limb of the loop of Henle (DLH) (which is the continuation of the PCT) and the ascending limb of the loop of Henle (ALH) (which in itself has two parts, thin and thick; we will for the purposes of clarity refer to the descending and ascending limbs only). The descending limb of Henle is only permeable to H_2O, but not permeable to solutes. This means that only water will be able to flow through here. However, keep in mind from our previous discussions that in order to move water, you must first generate a gradient for the movement of water! If you look at Figure 5.8, you'll notice that while tubular fluid is flowing in one direction, the blood flowing through the peritubular capillaries flows in the opposite direction. This increases the peritubular capillary concentration of NaCl that is being reabsorbed in the TALH. As the ALH is not permeable to water, there will be an increase in the peritubular capillary's osmolality. When the peritubular capillary comes into touch with the descending limb of Henle (which IS permeable to water), the water will flow into the peritubular capillary lumen following the osmolality gradient! Let's analyze a bit further how the salt is reabsorbed in the ALH.

The main Na^+ transport protein in the ALH is the $Na^+/K^+/2Cl^-$ or NKCC2. In Figure 5.9, we can see that NKCC2 reabsorbs Na^+, K^+, and Cl^- from the tubular lumen using the Na^+ gradient generated by the Na^+/K^+ ATPase. As it reabsorbs Na^+, K^+, and Cl^-, the K^+ is recycled to the lumen, thus making the lumen have a positive charge and the

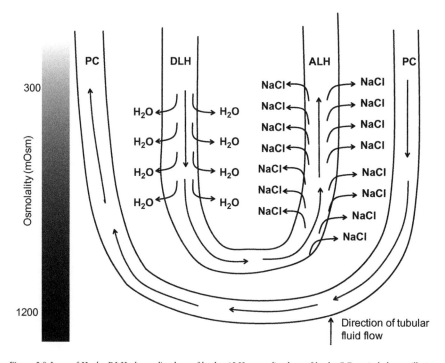

Figure 5.8 Loop of Henle. DLH, descending loop of henle; ALH, ascending loop of henle; PC, peritubular capillaries.

basolateral space negatively charged due to the reabsorption of two nega-tive charges (Cl⁻) and one positive charge (Na⁺). Because the cell is neg-ative, it favors the passive reabsorption of Ca^{2+} and Mg^{2+}. This transport protein can be stimulated by a number of hormones including antidiuretic hormone (ADH) and parathyroid hormone (PTH).

Clinical Correlate

Loop diuretics (furosemide and bumetanide) directly inhibit NKCC2, decreasing the reabsorption of Na⁺ and water, thus increasing urinary volumes. These are the most potent diuretics available in medicine and are excellent for treating conditions such as pulmonary edema (excess fluid in between the alveolus and capillary, thereby inhibiting O_2 exchange). However, these diuretics can generate grave alterations in elec-trolyte balance (hyponatremia, hypokalemia, hypocalcemia) if not used with caution.

So, when exactly does the body need the loop of Henle to function and begin reabsorbing water? When the ECF osmolarity increases

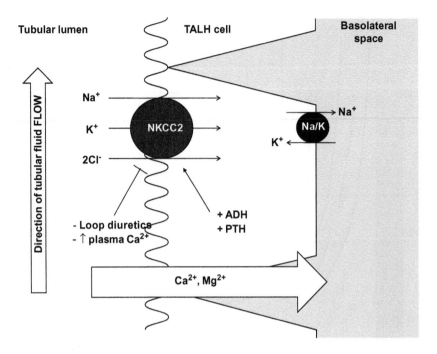

Figure 5.9 Loop of Henle.

either through a net gain of salt or loss of water (Chapter 1). When there is a net deficit of water, the body has a specific hormone that aides in concentrating urine and thus in conserving as much water as possible: ADH. ADH is produced by the hypothalamus and released by the posterior pituitary when ECF osmolality rises above 280 mOsm. How exactly does ADH work? The most straightforward way to look at it is Figure 5.10. The activation of NKCC2 by ADH will stimulate the concentration of NaCl in both the peritubular capillaries as outlined previously and in the renal interstitium, thus generating a water reabsorption gradient. ADH will concomitantly increase the permeability of water by opening water channels (aquaporins), which will allow for water to diffuse from inside the tubules to the interstitium and ultimately to the peritubular capillaries and blood. The sole release of ADH from the posterior pituitary favors the retention of FREE WATER! The take-home message from our previous discussion is that activation of NKCC2 leads to free water arrival in the distal portions of the nephron! Which means that if we want to keep that free water, we'll activate aquaporins to reabsorb it. Remember, if NKCC2 is activated, free water is generated!!!

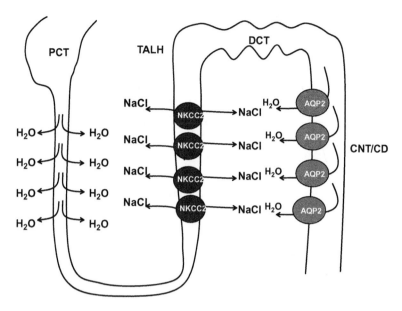

Figure 5.10 Actions of ADH. ADH/AVP will activate NKCC2 thereby concentrating the medullary interstitium. In addition, ADH/AVP will activate AQP2 which will reabsorb water following the salt gradient generated by the previous activation of NKCC2.

The reabsorption of Na^+ that occurs along the ALH is a key part in the functioning of the nephron. Not only will it regulate the reabsorption of other ions and water as we saw earlier, but it will help regulate the filtration rate itself! This is achieved through the Macula Densa (MD), in a process called tubuloglomerular feedback. If you look at Figure 5.11, you'll see an arrow that points at a shaded region of the ALH that sits between the afferent and efferent arterioles. This region is called the MD, and it will use Cl^- as an indirect measurement of filtration rate. This means that the arrival of a large amount of Cl^- to the MD will be interpreted as a high GFR and the arrival of a small amount of Cl^- will be interpreted as a low GFR. Now, from all our previous discussions, you've seen that the purpose of the body is not to increase or decrease randomly but to COMPENSATE in order to maintain homeostasis!! So if a decrease in the amount of Cl^- that's arriving to the MD is sensed as a DECREASE in GFR, the MD will signal the juxtaglomerular apparatus, JGA (a specialized group of cells located adjacent to both the MD and the glomerulus) (Figure 5.11), to increase GFR by promoting afferent arteriole vasodilation through the production of the proinflammatory mediators, prostaglandins. This means that if the MD senses a decrease in GFR, by

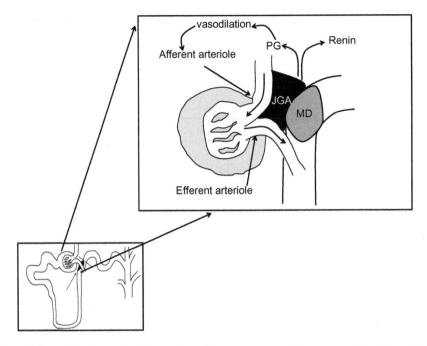

Figure 5.11 As Cl⁻ arrives to the MD (gray), it will sense the amount. A low amount will be interpreted as a direct marker of low glomerular filtration. This will signal the JGA (black), which will do two things: (1) Increase the production of prostaglandins (PG) to favor vasodilation of the afferent arteriole and increasing filtration. (2) Secretion of renin to stimulate the formation of Ang II.

dilating the afferent arteriole, there will be an increase in the hydrostatic pressure at the level of the glomerular capillaries, therefore favoring filtration! If there is an increase in the GFR through an increased arrival of Cl⁻, then the MD will signal the JGA to stop the production of the prostaglandins, thereby increasing R at the afferent arteriolar level and thus decreasing filtration. (This explanation is a concise oversimplification of the precise mechanism, the truth is we still don't know exactly how this takes place, but this is a summary of what is known thus far. If the reader is interested in a more detailed explanation, we recommend a more indepth physiology textbook).

5.2.3 A Brief Primer to the Renin−Angiotensin−Aldosterone System

We mentioned in the previous section that decreases in the arrival of NaCl to the MD stimulate both the production of prostaglandins (to increase the single nephron filtration rate) and the production of Renin. Renin is an enzyme produced by the JGA, and is the rate-limiting step in the formation of angiotensin II (Ang II) — one of the key hormones that regulate both acute and chronic changes in blood pressure through its action on the blood vessels and volume retention by the kidney respectively. As seen in Figure 5.12, renin will cleave a precursor peptide called angiotensinogen into Ang I. In turn, the angiotensin converting enzyme (ACE) will cleave Ang I and produce Ang II! It is Ang II that will have acute effects (vasoconstriction) and chronic effects, such as increasing volume handled by the kidney and increased production of aldosterone by the adrenal gland. With this in mind, let's now turn our attention to the final portion of the nephron where Ang II and aldosterone have a major effect!

5.2.4 The Distal Nephron and the "Fine-Tuning of Urine"

After ALH and MD comes the most distal part of the nephron which reabsorbs approximately 10% of the glomerular filtrate. It is composed of the DCT and the CNT/CD. Together the expression of a particular set of transport proteins along these cells will determine the final composition of urine and of plasma! (By regulating what actually ends up in the urine, the distal nephron will regulate what ends up in the plasma as well; therefore, the role of the distal nephron in regulating the fluid and electrolyte composition of the body can't be underestimated.)

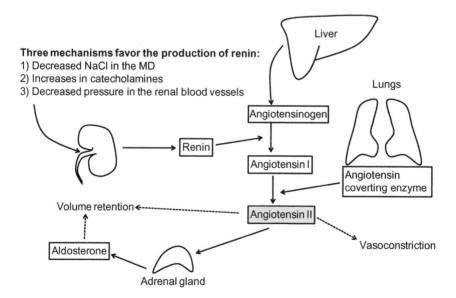

Figure 5.12 The activation of the renin–angiotensin–aldosterone system.

The first portion of the distal nephron is the DCT (Figure 5.13). This is going to be one of the MOST important regions of the nephron with regard to volume and electrolyte regulation. A transporter called NCC (NaCl cotransporter) is expressed at the lumen and reabsorbs NaCl, a.k.a. TABLE SALT!!! (Remember our conversations regarding salt and water from the first chapters?) By activating this transporter, the body will reabsorb salt, this salt will in turn pull water and this concomitant reabsorption of BOTH salt and water will lead to isotonic fluid retention! (If you go back to Chapter 1, you'll see that isotonic fluid is freely distributed in the EC and is thus KEY in repleting the intravascular volume.) The NCC is inhibited by a group of diuretics called thiazides.

Clinical Correlate

Arterial hypertension currently affects 1/3 of Americans. The first line of treatment according to the current guidelines is a thiazide diuretic! This underlines that chronic retention of salt and water is believed to be key in increasing blood pressure and thus play a major causative role in the development of arterial hypertension.

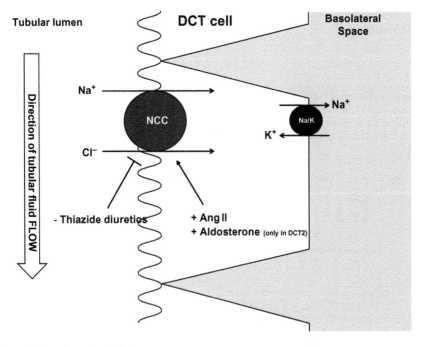

Figure 5.13 Distal convoluted tubule.

The key thing to remember about NCC is that it reabsorbs SALT and thus isotonic fluid which replenishes the intravascular volume. In tune with the reabsorption of salt, NCC is activated by the two quintessential volume deficit hormones: Ang II and aldosterone!! This means that when the body lacks volume, it will produce Ang II and aldosterone, which will, among other things, activate NCC to favor volume repletion!

The more distal part of the nephron includes CNT/CD. It is in this part of the nephron that a very particular type of transport takes place. Via two channels, one for sodium (ENaC, epithelial Na^+ channel) and one for potassium Renal Outer Medullary K^+ Channel (ROMK), Na^+ is essentially "exchanged" for K^+ (Figures 5.14 and 5.15). The reabsorption of Na^+ all by itself leaves a negative charge in the tubular lumen. If the tubular lumen is negative, this will "attract" (sort of like a magnet) positive charges from where it can. Because inside the cells there is a high concentration of K^+, this K^+ will respond to the call of the negative charges and it will be secreted to the tubular lumen via ROMK; thus we can say that in this part of the nephron, Na^+ is actively being exchanged for K^+.

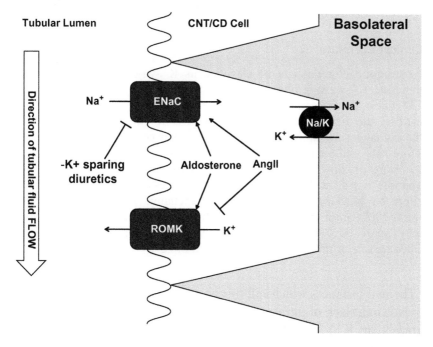

Figure 5.14 Connecting tubule/collecting duct.

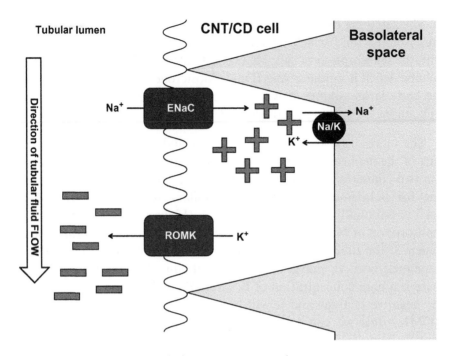

Figure 5.15 ENaC/ROMK mediated Na⁺/K⁺ exchange mechanism. Na⁺ entering through ENaC will generate a lumen negative voltage and a cell positive voltage. These voltage differences favor K⁺ secretion via ROMK.

Clinical Correlate

Nonosmotic diuretics can be classified as K^+-wasting and K^+-sparing diuretics. K^+-wasting diuretics have a target before reaching the CNT/CD (loop diuretics and thiazides) by blocking reabsorption of Na^+ increase the arrival of Na^+ to the CNT/CD K^1 wasting diuretics favor the exchange mechanism of Na^+ and K^+ between ENaC and ROMK. More Na^+ will get to the CNT/CD because it is not being reabsorbed in the more proximal segments; therefore, more K^+ will be secreted. The opposite is true with K^+-sparing diuretics, these diuretics have a direct effect on blocking the ENaC/ROMK exchange mechanism and thus waste Na^+ (hence their diuretic effect) but can no longer secrete K^+ (thus they "spare" K^+). In patients with hyperkalemia, loop diuretics and thiazides are one of the options for getting rid of K^+!

The two instances which call this portion of the nephron to action are hypovolemia (lack of intravascular fluid) and hyperkalemia (an excess of extracellular K^+). During hypovolemia, Ang II and aldosterone are secreted to prevent further losses of Na^+ and water, thus favoring the reabsorption of Na^+ all along the nephron, but Ang II inhibits the secretion of K^+, thus the ultimate effect of Ang II is to retain volume! During hyperkalemia, however, aldosterone is secreted alone (no Ang II); the sole secretion of aldosterone will favor the ENaC/ROMK mediated exchange of Na^+ for K^+, thus excreting K^+ into the urine.

The most distal part of the nephron is important when dealing with "acid−base physiology." An in-depth discussion of the mechanism that are behind total body acid−base regulation are beyond the scope of this book; however, we include the following nuggets of information which we think are useful in understanding the general concepts behind the kidney's handling of acids and bases:

1. The body always tends toward acidity, we produce H^+ ions daily that need to be excreted.
2. The PCT is key in reabsorbing and generating HCO_3^- (this means that the filtered bicarbonate will be reabsorbed).
3. The distal nephron (CNT/CD) is key in generating "new" HCO_3^- (this means that the bicarbonate will be generated by the cells by the action of carbonic anhydrase and the H^+ will be buffered in the tubular fluid, thus generating "new" HCO_3^- that will then be reabsorbed).
4. H^+ will be actively secreted into the tubular lumen by the cells in the CNT/CD through two mechanisms (Figure 5.16):
   a. The H^+ pump which pumps protons into the lumen.
   b. The H^+/K^+ exchanger.

We have now finished our "brief" discussion on the nephron! We have discussed molecular mechanisms further beyond what we did in previous chapters because we believe that understanding, why and how, the kidney is able to concentrate urine, retain volume, and secrete potassium through the action of different transporters, is important in understanding the clinical repercussions of the different physiologic states that we will discuss in the following chapters.

The take-home message for nephron function is highlighted in Figure 5.17. Each portion of the nephron will have a distinct contribution to the final makeup of both the blood and the urine. If we understand that each of these mechanisms is interdependent, then we will begin to understand how complex this system really is. Through the next chapters, we will integrate the knowledge gained in our previous

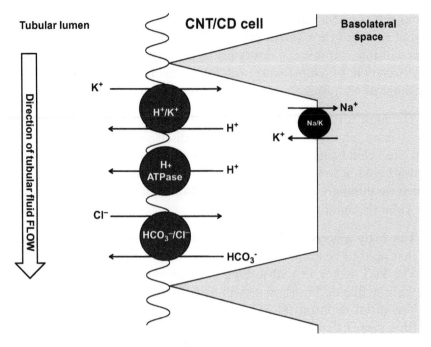

Figure 5.16 Transporter/exchanger/pumps along the connecting tubules and collecting ducts. This part of the nephron is key for regulating acid/base balance.

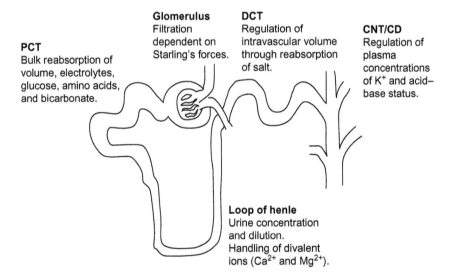

Figure 5.17 A summary of the multiple functions of the nephron.

chapters, in an attempt for the reader to comprehend how the body responds to fluid challenges.

5.3 CLINICAL VIGNETTES

A 50-year-old diabetic patient presents to clinic for a 6-month checkup. He reports that he has been doing well and taking all his medications as prescribed. He takes insulin before every meal, an ACE inhibitor in the morning, when he develops swelling he takes some furosemide (Lasix®) (a loop diuretic), and he takes metoprolol (a beta blocker) every day.

1. *The increased levels of glucose in plasma will lead to an increased filtered load of glucose. Glucose is freely filtered by the glomerulus and the PCT of the nephron is responsible for the reabsorption of glucose through the cotransport of Na^+ and glucose. The loose junctions between PCT cells reabsorb water, which follows the osmotic gradient generated by the reabsorption of Na^+, glucose, and other solutes. In this particular patient, what can happen if the glucose load exceeds the capacity of the PCT to reabsorb glucose?*
   A. The glucose will be reabsorbed in a more distal portion of the nephron.
   B. It is normal for glucose not to be reabsorbed in the PCT and ends up in the urine.
   C. The glucose will not be reabsorbed and will generate an osmotic gradient that will increase urinary volume.

   Answer C. The only place where glucose is reabsorbed in the nephron is the PCT. Saturation of glucose transport occurs when the plasma glucose concentration exceeds 180 mg/dL (normal <110 mg/dL). Whenever the glucose transporters in the PCT become saturated, glucose will inevitably end up in the urine because there are no other glucose transport proteins. Urinary volume will increase because the glucose that is not reabsorbed will lead to an increased osmotic gradient between the intratubular fluid (high in glucose) and the tubular cells. This will ultimately lead to an increased content of water in the urine. This explains the classic symptoms of polyuria (lots of urine) and the subsequent polydipsia (increased thirst) that is the hallmark of diabetes.

2. *The patient has some complaints though. He mentions that every time he takes furosemide (Lasix®), he has terrible cramps the*

following day. He has found that eating tomatoes, oranges, and bananas improves the cramping somewhat. He wants to know why the cramping is occurring.

A. The cramps are purely psychological and he needs urgent psychiatric attention.

B. The administration of furosemide leads to low glucose levels which are impacting the muscles' ability to contract, resulting in cramps.

C. The administration of furosemide results in hypokalemia and alterations in divalent ion handling which has been associated to cramping.

Answer C. The administration of diuretics increases the excretion of Na^+ and water by inhibiting Na^+ reabsorption along the nephron. The administration of furosemide, a loop diuretic, will inhibit NKCC2 and will increase the excretion of Na^+ and water, aiding getting rid of some of the edema this patient has. The administration of furosemide will increase arrival of Na^+ to the distal part of the nephron. This increased delivery of Na^+ to the DCT and the CNT/CD, where Na^+ is exchanged for K^+, the more Na^+ arrives to the distal nephron, the more K^+ that gets excreted. Additionally, the inhibition of NKCC2 leads to a decreased reabsorption of divalent cations such as Ca^{2+} and Mg^{2+}. The loss of these electrolytes is associated with cramping and thus the symptoms this patient is experiencing.

3. *The renin–angiotensin–aldosterone system plays a key role in the reabsorption of Na^+ and the response to decreased volume. The activation of this system is dependent on the production of renin. The MD is particularly sensitive to variations in the arrival of NaCl. A decrease in the arrival of NaCl will trigger the secretion of renin to favor an increase in GFR and stimulate the production of Ang II. The administration of an ACE inhibitor will have what effect on this patient's GFR?*

A. Increase

B. Decrease

C. Remain unchanged

Answer B. Ang II is produced when the intravascular volume is depleted. As such, intravascular depletion will be accompanied by a decrease in renal blood flow. Decreases in renal blood flow lead to decreased hydrostatic pressure within the glomerular capillaries, which will decrease the GFR. Ang II however preferentially vasoconstricts the efferent arteriole and aids in the production of nitric oxide in the afferent side, thereby increasing the GFR even in the presence of decreased GCH. Additionally, the renal retention of Na^+ and water will lead to increased blood pressure, increased perfusion of the kidney, and increased GFR. The administration of inhibitors of the ACE will decrease GFR because it opposes the aforementioned effects. However, in patients with hypertension, the decrease in blood pressure and the concurrent protection of the renal corpuscles by the administration of an ACE inhibitor is a major therapeutic objective.

CHAPTER 6

Integrated Response to the Loss of Blood

The only way to understand the body's response to a derangement in homeostasis, that is, a diseased state, is in an integrated manner. Each system will respond in its own particular way, and it's understanding the way these individual responses fit together that help doctors assess and treat any problems at the bedside.

Let us begin with the body's response to loss of blood. Let us be very specific, we are talking about the loss of blood, not a loss of hypotonic or hypertonic fluid, but whole blood. Remember from Chapter 1 that loss of isotonic fluid does not alter the distribution between extracellular/intracellular (ECF/ICF), nor does it cause sudden shifts between intravascular compartment and the space (IV/IT); it only leads to a decrease in ECF volume. Remember that in order to shift between the ECF and the ICF, we need to change osmolality, and in order to shift between the IV and IT we need to alter Starling's forces and in particular capillary oncotic pressure. This means that by losing blood, initially ALL we are doing is depleting the blood vessels of some fluid!

Let's look at this a little more closely. What is the primary goal of blood. (Remember Chapter 4?) To provide oxygen to the tissues! Blood evolved as a tissue that was able to increase the delivery and exchange rate of oxygen between the external environment and the internal cells that require the oxygen to create energy more efficiently. In order for blood to carry out its duties, the cardiovascular (CV) system needs to maintain adequate blood pressure (BP)! Out of all the organs, there are two organs that are the most "oxygen hungry." What two do you think they are? Your brain and your heart! Well before these organs start being seriously threatened by a lack of oxygen, the body will go into panic mode and will activate any and all necessary mechanisms to keep oxygen delivery to these two organs stable, even if this means shutting down the rest of the body.

Now before we get too far ahead of ourselves, let us explain that during blood loss, the goal of the body is to maintain perfusion of the heart and brain by maintaining BP! For the purposes of simplicity, we can say that the body's response to the loss of blood can be divided into two categories. These are: (1) acute and (2) chronic responses. We can think of an acute response as one that does not correct the underlying problem; rather, it simply corrects a symptom. With respect to blood loss, an acute response will attempt to maintain BP, while a chronic response will attempt to recover lost volume. BP is the amount of pressure that the arterial system is exerting, and the ultimate goal is to maintain target organ perfusion.

Key

Target organs are the brain and the heart.

BP has two main variables, cardiac output (CO) and total peripheral resistance (TPR) ($BP = CO \times TPR$). CO is the amount of blood the heart pumps in a minute; it is expressed in L/min. TPR is the added resistance of all the arterioles in the body.

Key

If the arterioles impede flow, this will cause the pressure in the upstream segment (the arterial system) to increase. Thus greater flow will go to the places with the least amount of R, that is, the heart and brain!

So if you lose volume, the key issue is to maintain the perfusion of the brain and the heart. So in an ideal situation, you would want to recover the lost volume ASAP; however if this is not possible, your body has to find a way to maintain BP in the face of decreased volume, in order to maintain perfusion of the heart and brain. How would you do this? **You will maintain BP through increasing CO, TPR or both simultaneously!!!** The aim of the acute responses will be to eliminate "the sign," that is, low BP in spite of not correcting the problem which is low volume. **Most often, the first sign you see clinically is an increase in heart rate!** A chronic response would involve correcting the problem, so a chronic response will have an impact on fluid retention! (Remember we are trying to replace blood.)

Key

If $BP = CO \times TPR$, to increase BP we have to:

1. Increase CO
   and/or
2. Increase TPR.

So let's see how this would play out in real life. A perfectly healthy 25-year-old, 70 kg man walks into the blood bank in order to donate blood. He talks to the nurse, fills out the paperwork, and gets his vital signs taken. He has a heart rate (HR) of 70 beats/min (normal between 60 and 100), BP of 110 mmHg systolic, and 70 mmHg diastolic (normal is roughly 110/70 mmHg) and breathing rate of 12 per minute, which is also normal. He sits up on the blood donation couch, stretches out his arm and the nurse inserts a big needle into his arm, blood starts to pour out slowly through the tubing into a special reservoir bag, this continues until he has lost approximately 500 ml of blood (which represents approximately 10% of his circulating volume!). Because the blood loss is isotonic with the rest of the body, there will be no fluid shifts between ICF and ECF, and more importantly between interstitial and IV, at least not at first. If there are no fluid shifts, this means that the only compartment that will be depleted is the IV. If the IV is depleted of volume, then the pressure within it will decrease! This leads to the presentation of our problem-sign binomial. Problem = 500 ml volume depletion, sign = low BP. Throughout the rest of the chapter, we will refer to mechanisms that only attempt to correct BP as acute and to the mechanisms that correct volume depletion as chronic. **(Word of warning, BP is only decreased clinically when there's a volume deficit of 30–40%; however, the sensing mechanisms for decreased BP do involve sensing the small minute-to-minute variations in BP which are physiologically—albeit not clinically—significant.)** In addition, it is important for you to consider how different organs in the body work TOGETHER in trying to achieve a common goal. All of these responses are geared toward the same thing: PERFUSING THE BRAIN AND THE HEART WHILE RETAINING FLUID. Keep this is mind!

6.1 THE ACUTE RESPONSE

6.1.1 Role of the Heart, Blood Vessels, and the Other Organs in Maintaining BP

6.1.1.1 Increase TPR

Right, so now our patient has lost 500 ml of blood. . ... What's going to happen next? The first organ to respond is the brain! This is achieved through the baroreceptor mechanism, which acts as a sensor. In order to correct something that's gone awry, we need to sense it! When there are slight minute-to-minute decreases in BP, the baroreceptors in the arteries of the neck/chest (located at the carotid bifurcation) will activate the vasomotor center in the hypothalamus. Decreases in BP will activate the baroreceptor reflex and the vasomotor center itself will increase the sympathetic tone in the entire body (Figure 6.1), this means that throughout the entire body there will be an increase in the amount of circulating norepinephrine and epinephrine (NE and E). This increase in catecholamine secretion will stimulate *alpha* receptors

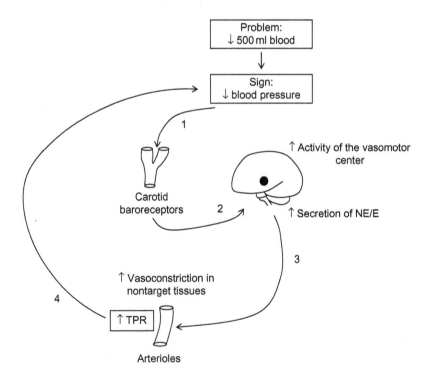

Figure 6.1 *The initial volume depletion stimulates the carotid baroreceptors so that there is increased sympathetic output from the vasomotor center in the brain. This in turn increases sympathetic tone, causing vasoconstriction in the arterioles. This increases TPR and thus increases BP, whithout changing circulating volume.*

in the peripheral tissues, particularly in the abdominal viscera and skin to generate vasoconstriction. The increased vasoconstriction increases resistance in nontarget organs and redistributes flow to more important tissues that have less abundant alpha-receptor expression. The brain and the heart have much fewer *alpha* receptors than other parts of the body. So you see, the response is hardwired!

This means that there will be increased resistance in areas that are not considered nonessential; for example, the body can make do with relatively very little blood flow to the intestinal, splenic, and skin circulation for a while. This concept is called redistribution of flow, because flow is redistributed to the critical areas (heart, brain, and initially to the muscle) and diverted from the less important areas (skin, spleen, intestines). Interestingly enough, the kidneys, which we might add are pretty important in responding to volume depletion, undergo vasoconstriction just like the intestine and the spleen. They have these *alpha* receptors as well. When the messages start getting sent out, that nonessential organs need to start clamping down on their blood supply to limit flow to nonessential parts, the kidney does so as well. Why? Well, this serves two purposes. For starters, the kidney normally receives a little more than 1/5 of the total CO (a little over 1 l/min). Clamping down on the amount of blood flow allows a large amount of blood flow to be redistributed to the heart and brain. Secondly, if you're volume depleted, you would like for the organ that is actually going to handle volume retention to know that volume is lacking! Much like a lieutenant fighting alongside his men to know what he's sending them into, the kidney in essence is right there "in the trenches" with the nonessential organs. It feels what they feel and can therefore respond appropriately!

In addition to the secretion of catecholamines and angiotensin II (Ang II) in the acute (BP correcting) setting, ADH is also secreted. ADH, also called vasopressin, in addition to its water-retaining effects which we saw previously, is a potent vasopressor (hence the name). It is released by the posterior pituitary of the brain as part of the acute response. It, like Ang II, serves to bridge the gap between acute and chronic responses. Acutely, it increases TPR like Ang II by acting directly on blood vessels (Figure 6.1).

Now, remembering what we learned in Chapters 2 and 3, blood vessel radius has a dramatic impact on BP! So you can lose volume, but this acute change in TPR means that despite a loss in volume (one part of the BP equation), a rapid decrease in blood vessel radius means that BP IS NOT going to be a DECREASED in the beginning! The rapid increase in resistance can initially make up for this loss of volume provided the loss is not too great. This is why a sustained drop in BP in an actively bleeding patient is an ominous sign that a patient's physiologic reserve is being stretched to the limit!

6.1.1.2 Increase CO
The second mechanism the body has for acute compensation of blood loss is to increase CO. How exactly are we going to increase CO? Well, in reality it is not so much an increase in CO as it is a "recovery" of CO, it's the body attempting to regain what's lost in terms of circulation. Remember the concept of mean systemic filling pressure and the vascular function curve? As we lose blood, venous return decreases because there is less blood filling our system! In Figure 6.2A, it is evident that by losing blood we shift the vascular function curve to the left because there will be a decrease in venous return, which in and of itself decreases CO. So, how exactly do we "recover" the lost CO?

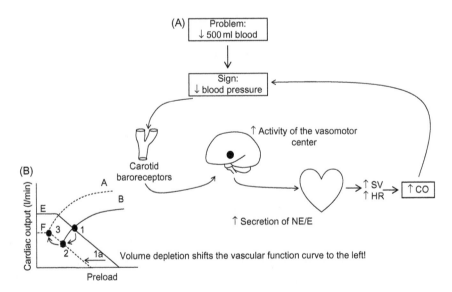

Figure 6.2 (A) Decreased BP leads to increased NE/E secretion, which will activate the beta receptors in the heart leading to increased SV and HR, these will increase CO and will thus increase BP. (B) Before volume Depletion, cardiac contractility and the vascular function curve are at the baseline (point 1), volume depletion initially shifts the vascular function curve to the left (1a) and the CO decreases (point 2). In order to recover the CO to normal, contractility must increase (through NE/E stimulation), increasing SV and HR thus recovers CO near baseline.

Through the catecholamine-mediated increases in contractility and heart rate! Catecholamines will stimulate the heart to beat harder (NE/E mediated Ca^{2+} release will cause an increase in contractility) increasing the stroke volume (SV). Not only do catecholamines improve the strength of each heartbeat, but they also cause the heart to beat faster, increasing the HR. If CO is the product of HR and SV and we increase both HR and SV, CO increases!

Simultaneous increases in TPR and CO increase BP, thus maintaining perfusion to heart and brain. In parallel to the response in the arterioles, the surge of catecholamines will also influence venoconstriction, that is, the constriction of the veins. Remember the vascular function curve from Chapter 3, where we stated that venous return = CO? Through catecholamine-mediated stimulation, the muscles in the larger veins leading toward the heart will contract. Contraction of the veins will favor increased venous return by decreasing the vascular volume of distribution. By increasing venous tone, we increase venous return and thus we increase CO, correcting our initial drop in BP (Figure 6.3). Be careful,

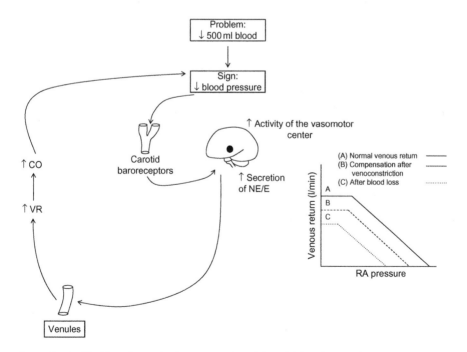

Figure 6.3 After blood loss, the vascular function curve (A), shifts to left (C), however after the increased veno-constriction will attempt to increase CO by shifting the curve back to the right (B) thereby increasing venous return and CO.

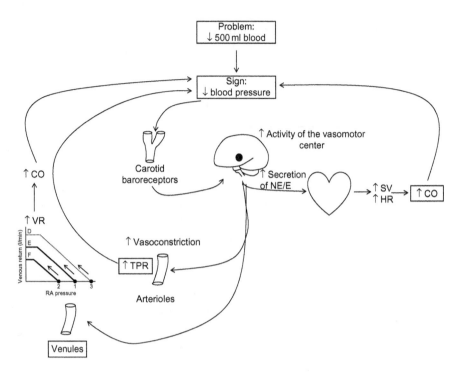

Figure 6.4 Coordinated response between the heart and the blood vessels in response to increased sympathetic stimulation.

venous vasoconstriction is NOT the same as arterial vasoconstriction! Arteries are built to handle pressure and veins are built to handle volume. As such, arterial vasoconstriction leads to increased resistance and consequently increases in arterial BP (immediate upstream segment). Veins however are made to handle volume and they function as a reservoir for blood. This means that blood is continuously flowing through veins but there is more blood than flow. If we squeeze veins, we squeeze reserve blood volume out of them, increasing venous return and thus increasing CO.

The coordinated "catecholamine" response is evident in Figure 6.4, where there are simultaneous increases in CO and TPR in order to correct the BP, in a manner that, at least in its initial phase is independent of increases in volume!

6.1.2 The Role of the Kidney in the Acute Response to Decrease in Circulating Blood Volume

Now, we mentioned that the kidneys are right there "in the trenches" with the rest of the "nonessential" organs in terms of response to catecholamines. The kidney also responds to these hormones acutely, so it can help by sacrificing some of its blood flow. However, it also plays a strong role in the formation of Ang II as mentioned in Chapter 5. Formation of Ang II is key in the body's response to loss of blood. Ang II will have effects both in correcting BP and increasing fluid retention. AngII is the hormone that bridges the response between acute (BP maintaining) and chronic (fluid-retaining) responses. We will start off with the acute response. Let us explain this in a little more detail.

The initial rise in catecholamines will favor prerenal vasoconstriction and thus a decrease in renal blood flow. This decrease in renal blood flow will have two effects. First, it will spare some of the blood flow so that a higher percentage of total blood flow can go to the heart and brain as we stated above. Secondly, because the vasoconstriction is prerenal (i.e., before the glomerulus), this will result in a decreased filtration rate. As less blood is filtered, this will stimulate the MD (by sensing decreased arrival of NaCl to the loop of Henle) to start producing renin. As you remember from Chapter 5, renin helps create the formation of Ang II by activating the liver precursor angiotensinogen. Angiotensinogen is normally hanging around waiting for this signal to convert itself into Ang I. This will in turn increase the production of Ang II as Ang I circulates through the lungs and comes in contact with angiotensin converting enzyme (ACE). Again, ACE converts Ang I to Ang II. Ang II will then stimulate multiple receptors throughout the body. These receptors are located in the heart and systemic and pulmonary arterioles (think back to the four points of resistance we mentioned in Chapter 2!) They are also located in the brain, kidneys, and adrenal glands. Acutely, they generate the following two key actions among a myriad of effects:

1. Increased TPR by generating arteriolar vasoconstriction by direct Ang II action.
2. Increased catecholamine secretion by Ang II acting directly on sympathetic nerve tone and on the medulla of the adrenal glands (which leads to more Ang II).

This second response is especially important to this process as it creates a positive feedback loop. Secreting catecholamines leads to increases in Ang II, which will then increase catecholamine secretion, etc. This feedback loop will only be broken when the problem (volume depletion) is corrected and the initial stimulus for formation of Ang II and NE/E stops.

Clinical Correlate

Clinically, you should note that there are additional methods by which renin is stimulated: beta-receptor stimulation in the kidneys by catecholamines and a baroreceptor reflex in preglomerular blood vessels which will sense decreased perfusion pressure (Figure 6.5). In total, the three mechanisms that stimulate the secretion or renin are:

1. Decreased arrival of NaCl to the macula densa
2. Beta catecholamine receptor stimulation in the kidney
3. Baroreceptor reflex in preglomerular blood vessels.

Over time, these effects can lead to a wearing down and an eventual remodeling of the heart to accommodate these changes, especially after the heart is injured such as in the setting of surviving a heart attack (myocardial infarction or MI). This remodeling is called neurohumoral remodeling. One of the more recent changes in the way we view management

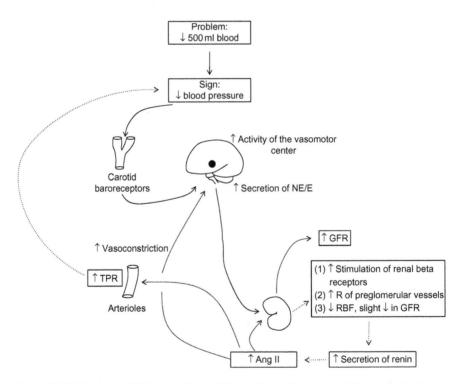

Figure 6.5 With the surge in NE/E triggered by the ↓ BP, the kidney will secrete renin (three mechanisms), renin will in turn produce Ang II which will further increase TPR and will activate the vasomotor center even more, leading to an increase in NE and E. Ang II will try to maintain GFR in spite of ↓ Renal Blood Flow (RBF) through predominantly efferent vasoconstriction.

of patients who have survived an MI is that stopping this neurohumoral remodeling is of paramount importance. If you look at antihypertensives available today, you'll see that there's a drug that attacks each one of these mechanisms. In the order of mechanism, we have diuretics, beta blockers, and ACE inhibitors/ARBs.

From what we've seen, the acute response (Figure 6.6) is aimed at one thing: maintaining BP. Note that in terms of the acute response, we have three major players: **catecholamines** (NE, E), **Ang II**, and **ADH/vasopressin** (there are others, but we will just focus on these three major ones). Catecholamines are entirely acute. They are rapidly released and are rapidly broken down. Part of the symptoms you see with patients suffering from acute blood loss is due to surge in catecholamines. You see tachycardia as *beta* adrenergic receptors in the heart are stimulated. You see pale skin as the blood is shunted away from the nonessential skin. You see anxiousness as receptors in the brain are

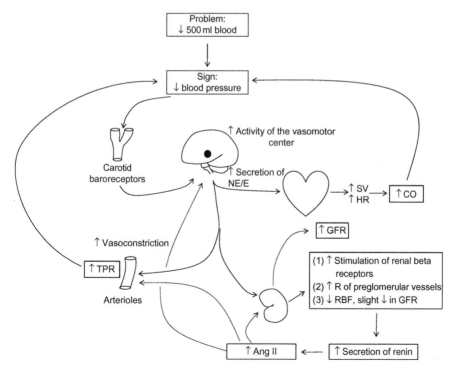

Figure 6.6 Integration of acute responses aimed at increasing BP without correcting volume status.

triggered, and you see decreased urine output as there is a global decrease in blood flow to the kidney.

Clinical Correlate

If you work in an ICU environment, sudden loss of fluids is met with resuscitation with isotonic fluid. If that doesn't work, and BP is still low, patients are often started on pressors (i.e., vasopressors). The most common ones used are catecholamines (e.g., NE, E, dopamine). Catecholamines act so quickly and are broken down so quickly that patients cannot be given these medicines intermittently. Instead, they are placed on "drips" which is essentially a continuous infusion. When you do your ICU rotations, you can see this in action and it's truly a sight to behold! Someone can have life-threatening hypotension, and you place them on a little norepinephrine and literally within seconds you can see the BP begin to improve. They are given as drips because, just as impressively, when you turn down the rate or stop them, you can see the BP drop within seconds.

Ang II and ADH−vasopressin also act acutely in addressing the symptom of a brief dip in BP. They act primarily as vasopressors and by promoting a positive feedback loop whereby catecholamines are continually released. However, unlike catecholamines, they also have chronic actions geared toward fixing not just the symptom but also the problem, as we will see shortly.

6.1.3 The Chronic Response to a Decrease in Circulating Blood Volume

The chronic response has two major goals:

1. Minimizing ongoing fluid/blood loss
2. Correcting the volume deficit.

Be careful though, minimizing ongoing loss and correction of volume deficit are not the same thing! Minimizing fluid/blood loss entails trying to stop active bleeding and excretion of fluid from the body; meanwhile, correcting the volume deficit involves a net gain of fluid. Let's further subdivide the two:

1. Minimizing ongoing fluid/blood loss
   a. *Activation of the coagulation cascade*—This is, at least in theory, relatively straightforward. If you plug the hole, you stop

bleeding. In reality, the coagulation cascade is incredibly complex and falls under both acute and chronic response. There is an initial platelet plug that tries to stop small vessel bleeding. This is more of an attempt to fix the symptom. The real benefit, however, comes over time as the coagulation cascade recruits additional cells to come and heal the damaged tissue by replacing it with intact/healthy tissue. Either way, the end result is minimizing ongoing blood loss.

b. *Reabsorption of as much NaCl and water from the glomerular filtrate as possible*—This will lead to concentrating the urine to minimize the loss of fluid while maintaining excretion of toxic substances.

In the first phase, minimizing losses, the goal is relatively straightforward: plug all leaks. Consequently the body will activate the coagulation cascade. Now, technically there are both acute and chronic aspects to "plugging the leaks." Initially platelets help form a primary plug which immediately treats the symptom (something you might notice if you've ever cut yourself with a kitchen knife). There is an entire self-propagating cascade of clotting that later takes place, but the specifics of the pathways are beyond the scope of this text. Later, these same pathways are used to invite new cells into the area to help fix the underlying problem by rebuilding the damaged tissue (think of the new skin that forms under the scab).

However, even if the initial leak is plugged, there are other losses to be considered. The most important one is urine! The production of urine is constant—so much so that an average adult will typically make about 1.5 L per day. This amount can vary widely as we learned in Chapter 5. But if you recall from Chapter 4, there is only about 5 L of blood in the human body! 1.5 L is almost 1/3 of your of your circulating blood volume, and it is on average about 2/3 of all the fluid you lose in a given day. Even in the face of volume loss, the kidneys need to continue to filter and rid the body of waste products. However, when burdened by more significant volume loss, the kidneys need to find the perfect balance between metabolic waste and water retention. The kidney works with a hierarchy; this means it will first address volume loss above all else. Only when the volume loss has been corrected will the kidney focus on correcting other abnormalities such as electrolyte and acid—base derangements. Now, the best way to retain volume (thus minimizing the loss of fluid through the urine) and excrete waste simultaneously is by concentrating urine! That way you excrete a high amount of waste in a small amount of water! By increasing the function of NaCl transporters and water channels (Chapter 5), the kidney will effectively retain as much NaCl and H_2O as possible, yet still excrete metabolic waste. In this way, it will minimize the amount of fluid lost, and in essence, partially plug one of the holes in our leaky container. This addresses the problem, by minimizing ongoing losses. It does not, however, do anything about replacing what's been lost, yet...

2. Correcting volume deficit
   a. Restoring the lost water and solutes (Ang II, aldosterone, and ADH)
   b. Restoring the lost cells and proteins (erythropoietin (EPO), etc.).

In order to correct a volume deficit, the body must not only retain and make efficient use of the volume it has as we just discussed, but it needs NEW FLUID, NEW SOLUTES, and NEW CELLS, in order to REPLACE what has been lost. In our example, the body lost 500 ml of blood and therefore needs to REPLACE 500 ml of blood. How exactly does the body do that? It replaces the blood component-by-component, thereby leading to a "net gain" of what it had previously lost. Figure 6.8 delineates the chronic or volume-replacing mechanisms the body activates in order to correct our problem. The focus has now shifted from BP, which in theory is stable, thanks to our compensatory mechanisms seen in Figure 6.7.

6.1.3.1 Restoring Water and Solutes: Role of ADH, Ang II, and Aldosterone

We stated early that acutely, ADH acts as a vasopressor to increase TPR. This serves as to treat the symptom of low BP. Over the longer term, ADH attempts to fix the problem through increasing free water reabsorption.

Chronic effects of ADH include the following:

1. Stimulating thirst. The best way to get more water is to drink it!
2. Stimulating free water reabsorption in the collecting duct.

Think of it like this: water has already been somewhat concentrated by the time it reaches the collecting duct. The collecting duct is the last chance to reabsorb water; therefore as we saw in Chapter 5, the salt gradient generated by the reabsorption of salt in the thick ascending

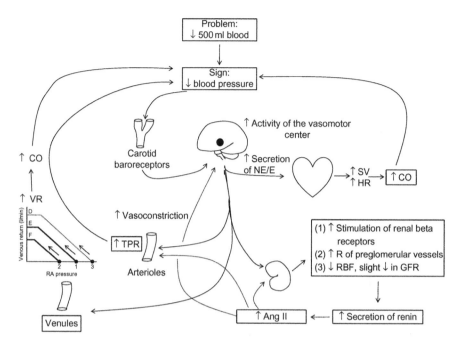

Figure 6.7 Integration of acute responses aimed at increasing BP including increased venous return.

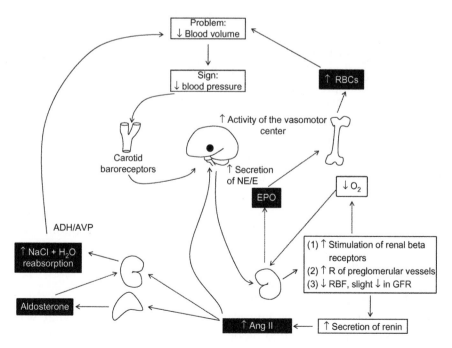

Figure 6.8 Integration of the chronic responses aimed at recovering the lost blood volume.

loop of Henle will help retain water as it passes through the collecting duct.

It should be noted though that ADH's role in isotonic volume or blood loss becomes somewhat of a neighbor who's trying to be helpful but can also prove somewhat annoying. It acts to reabsorb free water, but as we will discuss more in Chapter 7, we're not missing free water. We're missing isotonic solution! That is to say, we're also missing solutes! Thankfully, the body knows this, which is why it also makes Ang II/aldosterone.

Ang II, as we saw previously, has not only acute effects (vasoconstriction and increased NE/E release), but it also has chronic, volume-retaining effects.

Ang II has two main chronic effects:

1. Directly stimulates reabsorption of NaCl from the kidney,
2. Helps to stimulate secretion of aldosterone from the adrenal gland. Aldosterone then serves to increase reabsorption of NaCl in the kidney.

Together Ang II and aldosterone will "prime" the kidney to enter a volume-retaining state. Now wait a minute, we said that volume retention was a part of the response aimed at minimizing fluid loss, why are we mentioning it as a fluid gain mechanism? Well, imagine the following scenario (completely make-believe but humor us), let's say that you just lost your wallet and have no money. If all of a sudden someone starts tossing cash out the window, and you had no idea this was going to happen, you will try your best to grab a hold of as many bills as possible with your hands and place the money in your pockets. However, if I were to let you know that someone was going to toss money out of a window, you would probably run get a bag, box, or whatever you could to make your "money harvesting" more effective. The same thing happens with the kidney and volume retention. The kidney gets ready to go "volume harvesting" as soon as any of the volume is replaced.

6.1.3.2 Restoring Cells and Protein
The water and electrolytes however are only a fraction of what was lost when the patient donated blood. In order to correct blood loss, the body must generate blood. Interestingly enough, the decreased arrival

of oxygen to the medulla of the kidney triggers an alert indicating that the oxygen carrying capacity has decreased, prompting the production of EPO. EPO will then stimulate the production of RBCs from the bone marrow and thus restore the RBCs that were lost over time. In parallel, the production of proteins, white blood cells, and coagulation factors is similarly increased leading the correction of our initial problem. All of these concerted responses will ultimately lead to the restoration of the blood that was initially lost.

Clinical Correlate

In many cases of severe chronic kidney failure, this EPO production of the kidney is damaged. Subsequently, without certain medicines, most of these patients would become anemic over time. This is why you'll often see nephrology patients receiving EPO subcutaneous injections at least once a week.

Hemorrhage is a very common clinical problem dealt with by all surgeons, especially trauma surgeons. By looking at the American College of Surgeons Classification of Hemorrhage (Table 6.1), you can see that the clinical picture and the reasons why, should be more intuitive after reading this and the prior chapter. The clinical manifestations of the degree of blood loss correlate to the level of activation of catecholamine/Ang II surge. A small amount of blood loss will only lead to minimal changes including catecholamine-induced anxiety. As the blood loss becomes more severe, the myriad of effects of the catecholamine/Ang II increases will be evident, including pale cold skin (fluid redistribution), increased HR (to increase CO), decreased urinary output (renal vasoconstriction), and increased anxiety. In spite of ongoing

Table 6.1 American College of Surgeons Classification of Hemorrhage Severity

	Class I	Class II	Class III	Class IV
Blood loss (ml)	<750	750–1500	1500–2000	>2000
Heart rate	<100	>100	>120	>140
Blood pressure	Normal	Normal	Decreased	Decreased
Respiratory rate	Normal	Decreased	Decreased	Decreased
Urine output (ml/h)	>30	20–30	5–15	None/minimal
Mental status	Slightly anxious	Mildly anxious	Confused, anxious	Lethargic

Source: *American College of Surgeons Committee on Trauma – ATLS Student Course Manual 8th Edition.*

blood loss, BP remains normal even with a deficit that can reach almost 30% of blood volume!! This happens because of the acute mechanisms that are in play (Figure 6.6) to maintain perfusion to heart and brain. What is probably most important to take away from this chapter is that only after the mechanisms described above have failed, do we start to see a drop in BP! Thus, hypotension in the setting of blood loss is **always** a late finding, and it is **always** an ominous one! What is more, thorough understanding of the physiology of blood loss and the clinical repercussion is essential in order to treat it appropriately!

CC

The American College of Surgeons Classification of Hemorrhage involves four types, which are identified by % of blood loss: Type I < 15%, Type II 15–30%, Type III 30–40%, and Type IV > 40%. It is only after a patient has lost more than 30% of his/her blood that we are able to see a drop in BP.

6.2 CLINICAL VIGNETTES

An 18-year-old male is brought to the emergency department with a stab wound through his right thigh. He is bleeding profusely and is found to have a BP of 90/40 mmHg (normal >110/ > 70). His skin is pale and cool, his HR is 135. He is agitated and confused, yelling at the medical staff, but is not able to articulate words. A catheter is placed in his bladder and only 100 ml of concentrated urine are obtained.

1. *This patient is clearly suffering from massive blood loss. The tachycardia and low BP with a recent history of blood loss are all tell-tale signs that point to major intravascular depletion. The increased catecholaminergic tone found in this patient is responsible for the increased HR and at least in part for the agitation. In this setting, perfusion to the kidneys is:*
   A. Increased
   B. Decreased
   C. Unchanged
   Answer B. The kidneys are one of the most sensitive organs in the body when it comes to variations in BP. The increased level of circulating catecholamines will immediately provoke renal vasoconstriction. Renal vasoconstriction will lead to a decreased perfusion

of the kidney which is key in activating the response to blood loss. In order to minimize losses, the kidney will concentrate urine as much as possible in order to prevent further losses and to replenish the intravascular space as soon as repletion by mouth or in the form of intravenous fluids begins.

2. *The body's initial goal when faced with blood loss in the maintenance of arterial BP to favor perfusion of the brain and heart. An increase in BP can be obtained by either increasing _____ or _____.*

   A. CO, venodilation
   B. Venodilation and arteriolar vasoconstriction
   C. CO, arteriolar vasoconstriction

   Answer C. The formula for BP is: $BP = CO \times TPR$. By increasing either CO or TPR, or both, there will be an increased hydrostatic pressure within the arterial system, which will then redistribute blood flow to the target organs, that is, brain and heart. These compensatory mechanisms are quite efficient. However, as in the case outlined above, when the body's ability to compensate blood loss is overwhelmed, for example, a drop in BP to 90/40 is an ominous indicator that the intravascular volume is severely depleted and needs to be corrected urgently.

3. *In the case presented above, the clinical picture of hypovolemic shock is evident. In this case, in spite of increased venoconstriction, HR and arteriolar vasoconstriction BP is still low, thus compromising perfusion to brain and heart. After admission to the emergency department, the patient gets blood transfusions and isotonic crystalloid infusions which stabilize his BP somewhat. In the following days, he improves from the CV standpoint, he begins having dark bloody bowel movements and increased abdominal pain. What is the underlying cause of his gastrointestinal disorders?*

   A. A concomitant infection that damaged the intestinal and colonic mucosa.
   B. The blood transfusions generated a hypersensitivity reaction which leads to intestinal mucosa sloughing and bleeding.
   C. The intense vasoconstriction associated to hypovolemic shock led to intestinal ischemic damage.

   Answer C. As was mentioned previously, severe shock is associated to increased peripheral arteriolar resistance and shunting of blood to brain and heart. In the case of the intestines, when blood loss occurs, the intestines are one of the first organs to receive a decreased blood supply. If blood loss is not corrected emergently, the prolonged intestinal ischemia can lead among a number of things to mucosal damage and bleeding which can present itself as bloody bowel movements. Additionally, if the ischemic insult to the bowel is prolonged, sepsis and/or perforation can occur.

CHAPTER 7

Fluid Shifts in the Body!

The body has an amazing ability to adapt to change, and it is in understanding how the adaption process takes place that we can better comprehend the treatment strategies that we need to set forth in order to adequately treat our patients. In our previous chapter, we discussed the physiology of blood loss; however, we did not address specific changes in fluid (by fluid we mean aqueous solution of water, electrolytes, proteins, and other substances making up the extracellular fluid (ECF) and intracellular fluid (ICF) spaces, the intravascular (IV) and interstitial (IT) spaces). Blood loss is only one type of isotonic fluid loss. In our final chapter, we will briefly review the basic distribution disturbances seen in Chapter 1, but we will place more emphasis on clinical causes, compensatory mechanisms, and signs/symptoms associated to each one of the disturbances. In other words, we're going to tie together everything you've read into one wholly integrated clinical chapter.

The clinical impact of fluid alterations depends on the type of fluid that is being altered. Thus, we will classify all fluid alterations into one of the following six categories:

Isotonic disturbances
1. Isotonic fluid loss
2. Isotonic fluid gain

Hypotonic disturbances
3. Hypotonic fluid loss
4. Hypotonic fluid gain

Hypertonic disturbances
5. Hypertonic fluid loss
6. Hypertonic fluid gain.

As we mentioned in Chapter 1, the reader must understand that all fluid changes first occur in the extracellular (EC) space (Figure 7.1), and it is the changes in the EC space that can either generate changes in the intracellular (IC) space or they can limit themselves to altering

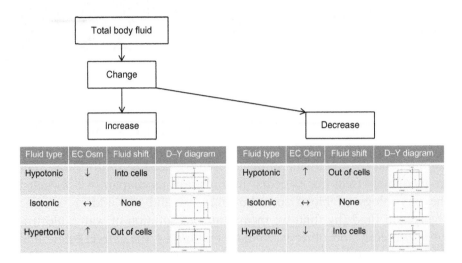

Figure 7.1 The six basic fluid alterations in the body.

only the EC space. Let us go over some brief concepts that we've previously described in Chapter 1:

1. Osmolality will define distribution of water between the EC and IC spaces.
2. EC hydrostatic and oncotic pressures will be major determinants of distribution of water between the IV and IT spaces.

We will now approach each of these scenarios from the clinical standpoint.

7.1 ISOTONIC DERANGEMENTS

In practice, there is no such thing as a purely isotonic change, with a purely isotonic compensation. Instead of black and white, we have many different shades of gray. We will, however, approach all the problems as black and white so that once we understand the "ideal" problem with an "ideal" solution, we can then mix and match to suit the needs of what is actually going on.

7.1.1 Isotonic Fluid Loss

The initial loss of isotonic fluid will decrease the total EC (the loss may be IT or IV), without altering the osmolality. This will not alter cell volume, as free water will NOT shift between ICF and ECF (Figure 7.2). However,

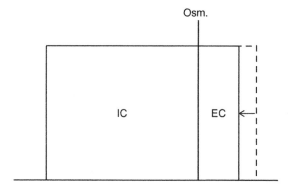

Figure 7.2 Isotonic fluid loss—EC volume contraction, with no overall change in total body Osm. By decreasing the TOTAL EC volume, we decrease both the IT and IV volumes!

the loss of fluid from the EC space will in most instances decrease total IV fluid volume and can thus compromise end organ perfusion.

Loss of isotonic ECF has several causes. Some of the most common are:

1. Blood loss (which is an entity all by itself and discussed as such in a previous chapter)
2. Diarrhea
3. Vomiting.

> **Key**
>
> Isotonic fluid loss leads to decreased ECF volume.

All the causes of isotonic fluid loss share the same basic theme. In each of the above cases, the proportion of electrolytes to water in the lost fluid is the same as that found within the cells. The type of electrolytes lost, however, will vary depending on the origin of the fluid loss. Let us explain.

Other than urine (which can be hypo-, iso-, or hypertonic depending on the situation), the body can produce *hypo*tonic fluid only in very few locations. These include the mouth, the colonic mucosa, and sweat glands. The rest of the body does not have the ability to concentrate or dilute fluid!!! This is an extremely important concept when it comes to fluid shifts! Remember that in order to move fluid in and out of the cell, we will need to change the ratio of water/solutes in the ECF vs the ICF. As most tissues are NOT able to concentrate solutes or fluid, most bodily fluids are ISOTONIC. This means that when we lose fluid, we're losing a balanced ratio of electrolytes-to-water; we're depleting the ECF stores, but the ICF volume stays the same. There will be individual concentration gradients affecting individual electrolytes, but the ECF/ICF osmotic gradient will be the same!

The loss of fluid from the EC space generally leads to a decrease in IV fluid. Why? Remember that in the distribution between IV and IT fluid, two factors are extremely important:

1. Oncotic pressure (water pull generated by the concentration of proteins)
2. Hydrostatic pressure (water push generated by the volume of water).

By decreasing the ECF, we decrease the volume of water and therefore the hydrostatic pressure in either the IV or the IT compartments. Provided that this loss is not equal on both sides (it usually isn't), this will lead to redistribution of fluid between the IV and IT spaces. Again, fluid will not redistribute intra/extracellularly, however. Let's analyze this on a case-by-case basis based on our common causes of isotonic fluid loss.

7.1.1.1 Diarrhea/Vomiting

Diarrhea can have many causes (infectious and noninfectious). Taking a simple approach we can say that the secretions in the stomach are isotonic and rich in hydrogen (H^+) and (Cl^-), while all post-pyloric gastrointestinal secretions are isotonic and rich in sodium (Na^+) and bicarbonate (HCO_3^-). This means that when we have vomiting, we will be losing HCl-rich fluid and when have diarrhea we will be depleting the ECF of bicarbonate-rich fluid. (This can ultimately wreak havoc on our acid–base balance, particularly if we are unable to replace the lost volume. We'll go into this a little later on.) A normal GI tract can secrete a huge amount of fluid each day (approximately 6–8 L per day!) However, it reabsorbs just as much or even more in order to maintain a steady-state volume. Vomiting is the expulsion of the contents of the stomach (sometimes even the contents of the intestines) by mouth. Diarrhea occurs when the reabsorption of this fluid is impaired or there is an increased secretion of GI fluid without an equivalent increase in reabsorption. The initial problem with vomiting/diarrhea is the immediate loss of fluid. Fluid loss in this instance can be complex, due to the fact that it's actually the parietal cells (HCl-producing stomach cells) or enterocytes (intestinal cells) that are losing what they've secreted into the GI lumen. However, this fluid loss DOES NOT count as ICF loss because the parietal cells/enterocytes are basically shuttling fluid between the GI tract and the IT space—acting sort of as the middle man who then exchanges fluid/solutes with the blood. **So the NET fluid loss is from the ECF and NOT the ICF**.

The loss of isotonic fluid will trigger the same salt and water retaining hypovolemic response that we saw in Chapter 6 in order to replace the lost volume. There are however two very important differences:

1. The fluid that we're losing, even though it is ISOTONIC, does not have the same concentration of Na^+, Cl^-, HCO_3^-, H^+, etc. as plasma. The response that is triggered in the kidney is aimed, at least initially, at recovering as much **NaCl and water** as possible. Once the initial hydration problem is resolved, the kidney will then start correcting acid/base and other electrolyte disturbances. These electrolyte disturbances can make you feel nauseous and cause further episodes of emesis. So adequate rehydration is the first and most important step in correcting the problem! Often times with rehydration alone, the kidney will correct the electrolytes on its own!

Key

Dehydration should be the first problem that is addressed before attempting to correct electrolyte or acid—base disorders with the exception of life-threatening hyperkalemia.

2. Oral intake or intestinal absorption may be impaired! In the setting of diarrhea/vomiting, the intestinal tract is not functioning properly. Vomiting and nausea can severely impair oral rehydration. Failure to tolerate oral rehydration can be fatal. Additionally, after extensive intestinal inflammation, the delicate absorption enzymes lining the small intestines can be damaged. This is particularly seen in pediatrics, where a toddler or infant will be able to take formula, but it will cause only further diarrhea. This means that if the ability of the GI tract to reabsorb fluid is compromised, intravenous fluids need to be considered!

The loss of isotonic fluid from vomiting/diarrhea is not just loss of sodium and water. The different types of solutes in the fluid depend on where in the GI tract the loss is occurring. However, they all lead to a clinical picture of dehydration, electrolyte, and acid−base disturbances. As we mentioned previously, the initial loss of isotonic fluid triggers the volume retention response by the kidney, which is aimed at reabsorbing as much NaCl and H_2O and consequently volume as possible, and unless the volume is replaced the kidney will continue to avidly reabsorb NaCl and H_2O, disregarding the fine-tuning of other electrolyte and acid−base disturbances as they begin to become abnormal. The fact of the matter is that the kidney needs a generous supply of NaCl delivered to the nephron in order to correct acid−base and other electrolyte disturbances. So, if it uses all of its sodium trying to drive water into the cells, then little will be left over for the acid−base and other electrolyte corrections that take place further down the pipe. (The specific transport paradigms are beyond the scope of this book; however, the reader should be particularly aware that the kidney requires the arrival of a sufficient amount of Na^+ to the distal nephron in order to adequately secrete H^+ and, likewise, the loss of Cl^- leads to an inability to secrete HCO_3^-). Ultimately, in the clinical setting, rather than focusing initially on correcting acid−base or electrolyte abnormalities, the primary focus should be placed on REHYDRATION. By doing so, the remainder of the electrolytes will almost invariably take care of themselves. The only caveat to that is with significant abnormalities in potassium.

Clinical Correlate

Elevated or low potassium (hyper/hypokalemia) can be a life-threatening emergency! You will learn in additional physiology classes, if you have not done so already, about cell membrane action potentials. In short, K^+ is essential for the adequate transmission of signals between cells. Of particular importance are the cells of the heart, the cardiomyocytes. Alterations in K^+ can lead to life-threatening arrhythmias! Thus attention must be paid to the K^+ levels in each and every patient which we are treating for volume depletion.

7.1.1.2 Impaired Oral Intake

One of the most common causes of pediatric death worldwide is dehydration due to infectious gastroenteritis. This can be due to inadequate oral intake or impaired absorption in the face of continued diarrhea and vomiting. Sadly, many, if not the majority, of deaths could be simply prevented by knowledge of oral rehydration. There are times when IV fluids are necessary (like, for example, when the child refuses to drink or has persistent vomiting). But as we mentioned above, getting some initial rehydration often allows the kidney to focus on correcting electrolytes and this causes vomiting and thus further oral intake to improve.

Clinical Correlate

The easiest rehydration strategy is always oral rehydration. If a patient is able to rehydrate by mouth, probably the best treatment to date is something called a reduced osmolality oral rehydration solution (ORS). It has an osmolality of about 245 mOsm and is the recommended method of oral rehydration endorsed by the World Health Organization (WHO). It is composed of the following:

NaCl—2.6 g/L or 45 mmol/L
Glucose—13.5 g/L or 75 mmol/L
KCl—1.5 g/L or 20 mmol/L
Trisodium citrate—2.9 g/L or 10 mmol/L.

There are actually commercially-available salt tablets than can be dissolved in water for use at home. These tablets save hundreds of thousands if not millions of lives each year in developing countries where there is less access to IV fluids and often more severe infectious diarrhea.

7.1.2 Isotonic Fluid Gain

Isotonic fluid gain (Figure 7.3) can be defined as a net positive fluid balance in the ECF compartment. As it is isotonic, it DOES NOT generate movement of water into and out of cells, and therefore it is almost invariably gain just in the ECF. Let's look at a common cause of isotonic fluid gain in a healthy patient. Probably the most common cause of isotonic fluid gain is iatrogenic, that is, caused by health-care professionals. A typical example would be giving a hospitalized patient too much of an isotonic IV fluid directly into their vascular space. If we're giving it in their IV, wouldn't that make the fluid gain primarily IV? Well, that depends! Recall from Chapter 1 that capillaries are permeable to water and electrolytes; thus any fluid and electrolytes that are pumped in the IV space will freely distribute in the **ENTIRE ECF**,

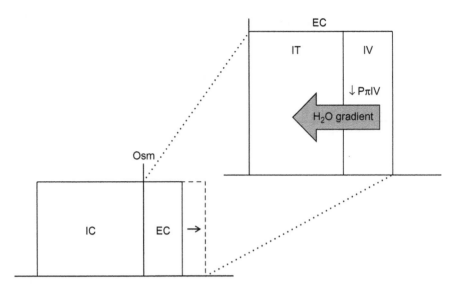

Figure 7.3 Isotonic fluid gain: (A) EC volume expansion, with no overall change in total body Osm. By increasing the TOTAL EC volume, we increase both the IT and IV volumes! (B) With decreased plasma oncotic pressure (PπIV), fluid leaks from the IV to the IT, increasing the IT space without altering IC volume.

including the IT space. Now let's examine a case of iatrogenic isotonic fluid gain in a little more detail to better understand what we mean.

7.1.2.1 Excessive Intravenous Fluid Administration

Let's say that a healthy 35-year-old female came to the hospital for a minor procedure. She's in otherwise good health and was NPO (nothing to eat or drink) the night before and the morning of the procedure. We place her on something called "maintenance IV fluids," which is basically the administration of isotonic saltwater through the vein. She has the operation and all goes well except that she lost a small amount of blood during the operation. However, we forget to turn off the fluids even when the patient starts eating. Thus she ends up with a good amount of excess fluid. In an otherwise healthy patient, the increase in IV compartment volume will cause a transient increase in hydrostatic pressure. Remember the carotid bifurcation that sense minute-to-minute dips in blood pressure which we addressed in Chapter 5? Well, there are different pressure receptors located in the walls of the heart that, when faced with elevated pressure, will produce a hormone called atrial natriuretic peptide (ANP) as well as B-type natriuretic peptide (BNP) as a response to increased distention of the cardiomyocytes. In contrast to what occurs during volume depletion with renin/Ang II and aldosterone, ANP/BNP *dilates* the renal arteries increasing overall blood flow to the kidneys! The kidneys receive this signal and begin secreting NaCl and H_2O into the urine to dump the excess fluid. Now, assuming the person is healthy and the amount and rate at which the fluid is given is modest, the heart and kidneys should be able to pee out most of this excess fluid without too much of a problem and without much increase in IT fluid. If the patient has a diseased heart or kidneys, the capacity to handle excess fluid can be significantly impaired and could even be life threatening.

Now, in the above example, we were dealing with a relatively healthy patient was given too much isotonic fluid. The clinical picture of isotonic fluid gain can change significantly in patients that have underlying disease processes. Even without knowing much about the following conditions, at first glance, you might not think that liver failure, nephrotic syndrome, and congestive heart failure (CHF) have a lot in common. However, there is a key characteristic that they all share: IV volume depletion with IT fluid excess. The reasons for the imbalance, however, are different.

7.1.2.1.1 Nephrotic Syndrome and Liver Failure
In both nephrotic syndrome patients, and patients with liver failure, there is a decreased plasma oncotic pressure as there is a decrease in circulating proteins, particularly albumin. In nephrotic syndrome, you have abnormally "porous" glomeruli and so you pee out protein when you're not supposed to. In liver failure, the liver is not able to produce enough albumin and patients suffer from decreased plasma oncotic pressure. The decreased plasma oncotic pressure leads to "leakage" of fluid from the IV to the IT compartment. Remember from Chapter 2 that IV oncotic pressure within the capillaries favors the maintenance of fluid intravascularly: it is the pull to the IV hydrostatic pressure's push. If the IV oncotic pressure decreases, fluid will leak from the IV to the IT compartment, decreasing the pressure within the IV space.

If the IV compartment does not have enough fluid, then those minute-to-minute low blood pressure receptors in the carotid bifurcation pick up on the decrease in pressure caused by the drop in IV volume. They then activate the hypovolemic NaCl and H_2O-retentive state we discussed in Chapter 6, mediated by catecholamines, Ang II, and aldosterone. This in turn leads to fluid retention because the body senses IV depletion. However, the truth of the matter is that there might be IV depletion, but there is a huge IT fluid expansion. In this instance, by trying to fix the symptom it does nothing to help the overall cause. It tries to absorb fluid when ideally it should be increasing oncotic pressure. Because of this, the fluid that the kidney is reabsorbing can't be maintained in the IV space and subsequently "leaks" into the IT space. As this isotonic fluid continues to move toward the IT space, this leads to the formation of generalized tissue edema! It often collects in gravity-dependent areas first, but eventually is evident everywhere! Because the hydrostatic pressure of the IT space has less effect on the Starling forces, massive amounts (liters upon liters) of fluid can diffuse into the IT space all the while maintaining inadequate IV volume.

7.1.2.2 Congestive Heart Failure

In patients with CHF, that is, a malfunctioning heart, the end result of generalized edema is the same. The IV oncotic pressure however is not reduced, but the cardiac output (CO) is! As CO is a determinant of BP ($CO = BP \times TPR$), the drop in CO can be equated to low circulating volume, in that the kidneys will activate the volume retention response and attempt to compensate the problem. The difference in this case being that when the problem is CHF (most commonly left ventricular systolic dysfunction), the kidneys are interpreting a state of lack of IV volume, when the actual problem is pump failure. This leads to fluid overload, which can be particularly dangerous in the lungs. The left ventricle has trouble dealing with this fluid and acts as an increased point of resistance to flow. Pressure backs up and of course just upstream from the left ventricle and left atrium are the lungs! The increased hydrostatic pressure in lung capillaries, which is caused by both pump failure and fluid overload, can dramatically decrease oxygen exchange as fluid builds up between where the oxygen in the air meets the blood in the capillaries. In essence, you can drown in your own fluids!

Clinical Correlate

Pulmonary edema is a worrisome complication of patients with CHF. The immediate goal of treatment is to increase oxygenation with supplemental oxygen and decreasing the fluid in the lungs with the use of loop diuretics such as furosemide (Lasix®), which will favor massive fluid loss.

The long-term treatment of CHF is aimed at minimizing chronic changes associated with the maintained activation of the catecholamine–Ang II–aldosterone response loop, which is why beta-blockers, ACEIs/ARBs, and spironolactone all play key roles in improving outcomes in patients with CHF. But one must be careful with indiscriminate use of diuretics in the setting of isotonic fluid gain. Remember that some of these conditions have LOW IV volume in the setting of HIGH IT volume. Giving a strong diuretic like furosemide could further deplete the IV space and actually cause organ damage secondary to poor perfusion.

7.2 HYPOTONIC AND HYPERTONIC DERANGEMENTS

The bulk of the essential knowledge regarding hypotonic and hypertonic challenges to the body which the specifics of fluid shifts between ICF and ECF was explained in Chapter 1. The major difference between hypotonic and hypertonic vs isotonic fluid changes is that hypo/hypertonic fluid shifts will have a DIRECT effect on cell volume. Most cells in the body have the ability to resist volume change by modifying their solute content. Neurons are the most vulnerable to changes in cell volume. Neuronal function depends on a stable cell volume. It is therefore logical that the predominant clinical features of hypo/hypertonic derangements are neurological! You must understand though that hypotonic and hypertonic fluid shifts can be fairly complicated to diagnose, evaluate, and treat. So we'll attempt to simplify things a bit by briefly discussing hypotonic and hypertonic derangements and focusing on the ECF Na^+ concentration and the compensation mechanisms in the body.

As we mentioned at the beginning of the book, Na^+ plays a key role in regulating fluid in the body. Being the most abundant ECF ion, **total Na^+ CONTENT determines total ECF VOLUME**. This means, because Na^+ is distributed in the ECF ONLY and it has the ability to draw water, the more Na^+ there is in the ECF, the more water there is, that is, ECF expansion. The CONCENTRATION of Na^+ has an effect not on total ECF volume, rather, it has an effect on how the fluid between ICF and ECF is distributed. In a steady state, there is a perfect balance between the Osm of ICF and ECF, so that there are no acute changes in cell volume. As Na^+ is the most abundant ECF osmolyte, most of the ECF Osm is dependent on Na^+. Osm is a measure of concentration, so if we change the concentration of Na^+, we will alter ECF osmolality and thus favor fluid shifts between ECF and ICF. Therefore, **cell volume is dependent on EC Na$^+$ CONCENTRATION**. As we will see, rather than present a problem with volume status, hypo/hypertonic fluid changes present a challenge to maintain ECF osmolality, that is, there is an alteration in total water content! Thus the body will approach the compensation of these variations somewhat differently, that is, instead of focusing on Na$^+$, in most instances, the body will generate variations in water content.

Key

ECF Na$^+$ CONTENT determines EC volume, while Na$^+$ CONCENTRATION determines IC volume.

7.2.1 Hypotonic Fluid Loss

Hypotonic fluid loss implies the loss of fluid with an effective osmolality less than that of plasma. As discussed in Chapter 1, losing this type of fluid leads to an increased osmolality of the ECF and subsequently to water diffusing out of the cells (Figure 7.4). Hypotonic fluid loss can happen in any of the following ways:

1. Profuse sweating after vigorous exercise
2. Increased insensible losses
3. Large amounts of dilute urine.

How does the body respond to this type of change? The simple answer is making sure that the body restores the lost salt and, more importantly, the lost water. The loss of a larger proportion of water,

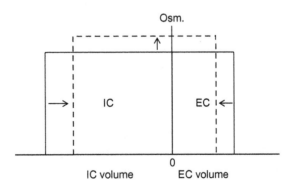

Figure 7.4 Hypotonic fluid loss.

which increases ECF osmolality, will set off a set of mechanisms in the hypothalamus, which ultimately lead to the secretion of ADH/AVP. Why? As we described in the previous chapters, the role of ADH in the kidney is to prevent the loss of water, pure and simple (pun intended). By increasing the secretion of ADH, the increased retention of free water by the kidney, which is achieved by concentrating urine (remember the loop of Henle reabsorption mechanism), will generate a positive balance of water without salt! This water will seep into the vascular space and then distribute itself throughout the body, correcting the initial loss of hypotonic fluid. Additionally, the body will increase thirst. This means that when your body's osmolality begins to rise, so does your thirst! This will combine two mechanisms: (1) renal retention of free water and (2) ingestion of free water, hopefully... This is if you actually choose to drink water and not a hyperosmolar beverage, which will only temporarily quench your thirst!

Clinical Correlate

Diabetes comes from ancient Greek *diabainein*, which means "to pass through, siphon." The old common name for it was "pissing evil" (bet you didn't learn that one in medical school!). There are two major forms: diabetes mellitus (DM) and diabetes insipidus (DI). *Mellitus* comes from Latin and literally means "honey." This is because ancient Greeks, Chinese, Egyptians, and Indians all apparently tasted the urine and found it to have a sweet taste. On the contrary, *insipidus* means "tasteless." Apparently, these patients' urine wasn't quite the same gustatory delight. Both of these terms, however, describe fairly accurately a major complication of both disorders: fluid loss. In both cases, the kidneys excrete large amounts of urine! DI occurs either because the brain doesn't secrete ADH (central DI) or because the kidneys lack the ability to recognize/act on it (nephrogenic DI). DM occurs because the elevated plasma glucose concentration overwhelms the ability of the kidney to reabsorb it. You will recall from Chapter 5 that ADH allows for concentration of urine (hence the name *anti*-diuretic hormone) and enhanced reabsorption of free water. If the hormone either isn't produced or doesn't work at the receiving end, then free water is not resorbed and rather it is dumped out into the urine. Therefore, the urine is "tasteless" because it is so dilute given all the free water. If not corrected, this will result in hypernatremia as the concentration of urine the ECF is increased secondary to loss of water > the loss of sodium. These patients will find themselves drinking massive amounts of liquids, even waking up 3−4 times per night to do so, but still they will not be able to keep up with the loss of free water, that is the power of the ADH hormone. In the case of DM, the glucose that is not reabsorbed will act as an osmotic diuretic. In both cases, there will be massive amounts of water lost in the urine, so if the patient is not able to keep up replacing the water by drinking or other medications, this will lead to dehydration!

7.2.2 Hypotonic Fluid Gain

Hypotonic fluid gain (Figure 7.5) is generally caused by consuming hypotonic fluid, whether it is by drinking a glass of water, by itself, or having an IV in your arm. A common clinical scenario for this is an inpatient setting in the hospital. Hospitalized patients are on all sorts of different IV fluids with different measures of water, salt, and sugar. These fluids, more often than not, are hypotonic. A prolonged infusion of hypotonic fluid will lead to decreased plasma osmolality! Remember that if the osmolality of cells is higher than that of the ECF, the cells will swell and whole body osmolality will decrease. Thus, a large increase in hypotonic fluid can generally have two immediate consequences depending on the velocity of infusion:

1. Expansion of the ECF. This will always happen, at least initially, no matter what type of fluid we are introducing to the body simply because we're adding volume; volume has to go somewhere. Although it will eventually be driven into the cells, the increase in ECF volume in the vessel will increase the hydrostatic pressure, decrease the oncotic pressure, and can result in edema as more fluid moves into the IT ECF and ICF spaces.

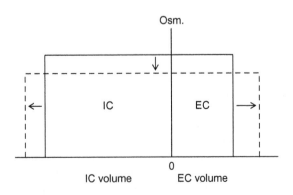

Figure 7.5 Hypotonic fluid gain.

2. Expansion of the ICF. As the ECF volume increases, the hypotonic fluid decreases ECF osmolality. This increases the ECF/ICF osmolal gradient and drives fluid into the cells until the ICF osmolality decreases to a similar amount. Clinically this can manifest itself as cerebral edema. Swelling of the neurons can give neurological symptoms such as headache, nausea, confusion, vomiting, seizures, and decreased levels of consciousness. The development of cerebral edema, however, is variable, depending on the amount of fluid, the rate of increase, and the ability of the neurons to compensate for the changes in ECF osmolality.

Compensatory mechanisms for hypotonic fluid gain and the two aforementioned complications are limited to producing large amounts of diluted urine and a regulatory volume decrease, a process through which cells rid themselves of solutes to decrease the ICF. Essentially, cells are able to shuttle ions, particularly Cl⁻, out of the cells in order to prevent swelling.

7.2.3 Hypertonic Fluid Gain and Loss

Hypertonic fluid alterations are generally only seen in the hospital setting. Hypertonic fluid gain (Figure 7.6) is generally secondary to specific medical treatments (infusion of hypertonic saline for resuscitation or management of cerebral edema), unless our hypothetical patient decided to swallow large amounts of seawater! In both of these scenarios, the general idea is the same: increased ECF osmolality and water movement from the ICF to the ECF compartment. The kidney, however, unlike with hypotonic fluid loss, has to deal with an increased solute load from the ingested fluid! This means that the kidney has to

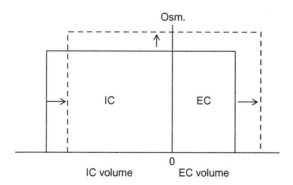

Figure 7.6 Hypertonic fluid gain.

necessarily excrete solutes into the urine in excess of what's coming in. Think about it this way, there is more salt than water being ingested! The problem that the kidney faces is that the maximum concentration of urine is approximately 1,200 mOsm/L, which means that any ingested fluid with an osmolality > 1,200 mOsm/L will generate unwanted water loss! Let us explain, seawater Osm approximates 2,400 mOsm/L. If we go and drink 1 L of saltwater, this means that we will have ingested 1 L of H_2O and 2,400 mOsm. The kidney however can only excrete 1,200 mOsm/L, so, the body needs to "give up" 1 L of its own water in order to dilute the saltwater and excrete all that extra salt that we took in! In the case of seawater, this will quickly lead to massive dehydration and consequently death! The administration of hypertonic fluids for resuscitation, although unlikely to provoke death from dehydration, must be done in a cautious and judicious manner to prevent rapid changes in cell volume.

Clinical Correlate

The administration of hypertonic fluids can be lifesaving in the setting of certain types of cerebral edema. Cerebral edema is so scary for two major reasons. One reason is, as we previously mentioned, neurons are very sensitive to rapid volume change. They lack the resiliency of other cells to bounce back from such changes. The second reason is where the cells are located: inside the skull! The compliance of the skull is about ... zero. Unlike the rest of the body, the skull will NOT distend in order to accommodate increased fluid volume. As the intracranial pressure begins to rise, the matter that exists within the skull will begin to be compressed. Eventually the brain will literally herniate, i.e., come out of the base of the skull. Unfortunately, one of the areas most susceptible to this is the brain stem, which is the center controlling of all of the most important life functions. By rapidly transfusing hypertonic saline solution (3% NaCl vs 0.9% isotonic saline), some of the fluid can be taken "shuttled out" of the intracranial space and delivered to the kidneys for excretion.

Hypertonic fluid loss is a normal occurrence when you concentrate urine. While this can be lifesaving if attempting to cross a desert, clinically it has little impact. This is because the mechanisms by which the kidney generates hypertonic fluid also lead to a small amount of urine output. Therefore, you don't lose very much water or overall solute relative to the total body solute content.

Based on our previous discussions, it is hard to separate volume, tonicity, and Na^+ derangements! Because sodium is the major contributor to osmolality, it is thus difficult to divide the discussion of hypotonic and hypertonic fluid shifts from the discussion of hypo- and hypernatremia as they are directly related. They are, however, NOT THE SAME! So we will briefly touch on hypernatremia and hyponatremia in an attempt to explain similarities and differences.

7.3 A BRIEF WORD ON HYPONATREMIA AND HYPERNATREMIA

Hypo- and hypernatremia are alterations in the measured plasma Na^+. This means that hypo- and hypernatremia refer to "Na^+ in the blood." This number however makes NO REFERENCE to the amount of fluid in the body, which is where to start to get dicey because both hypo- and hypernatremia can present with low, increased, or even normal plasma volume. What's even more confusing is that hypo- and hypernatremia are not actually disorders of Na^+ but disorders of water! Let us explain.

Hyponatremia is generally caused by an excess of free water, that is, water that has a lower EC concentration of sodium. A pure free water gain can generate cerebral edema, neurological impairment, and even death depending on the amount of excess water. This excess of water can be acute (minutes to hours) or chronic (days to months), which will greatly change how neurons adapt to these changes! If it is a rapid change, neurons will not be able to pump ions out fast enough to guard against an unhealthy ICF/ECF osmotic gradient. This generates cell swelling. However, if it is a gradual change, neurons are able to compensate which is why a Na^+ of 127 mEq/l that has developed over 3 months and in which the patient is asymptomatic is not as worrisome as a Na^+ of 133 mEq/l which developed in a matter of hours. However, hyponatremia can be accompanied by both normal, increased, or decreased osmolality and normal, increased, or decreased plasma volume!

Hypernatremia on the other hand is generally caused by an excessive loss of free water, that is, a loss of water with low Na^+, which concentrates the Na^+ in the ECF. Although hyponatremia is clinically more concerning in the acute setting than hypernatremia (owing to the risk of cerebral edema), hypernatremia is not devoid of risks due to neurological impairment. However, just like hyponatremia, hypernatremia can present with increased, decreased, or normal plasma volume!

Suffice it to say that a good rule of thumb for the treatment of both hypo- and hypernatremia that the correction of the Na^+ derangement should be done in a time frame approximate to the time it took for the initial disturbance to set in. This means rapid treatment for rapid onset and slow treatment for slow onset. As a comprehensive discussion of Na^+, alterations is beyond the scope of this book, and readers are referred to *Fluid, Electrolyte and Acid–Base Physiology*, 4th ed., by Mitchell L. Halperin et al. for an excellent case-by-case discussion on hypo- and hypernatremia.

Clinical Correlate

Plasma osmolality can be calculated (approximated) using the following formula:

$$\text{Plasma osmolality} = Na^+ \times 2 + \frac{\text{Glucose}}{18} + \frac{\text{Blood urea nitrogen}}{2.8}$$

This formula accounts for the most osmotically active components of the ECF, of which the most important is Na^+. So much so that a quick approximation of the osmolality can be achieved just by multiplying the Na^+ by 2! These three components (Na^+, glucose, and blood urea nitrogen) are not the only contributors to plasma osmolality; therefore, any changes in unmeasured osmolytes can lead to a difference between the calculated osmolality and the actual osmolality (which is measured in the lab), this is called the osmolal gap. The calculation of the osmolal gap can be used to identify toxins, as elevations in alcohols, sugars, lipids, and proteins which are the most common causes of an elevated osmolal gap.

7.4 PRIMER ON BEDSIDE FLUID MANAGEMENT

So, you may be asking yourself: are there practical applications to these concepts? Of course! Management and understanding of fluid distribution, movement, and alteration is key when taking care of patients in the hospital. It is key for several reasons. Fluid is the second-most immediate need of the body behind gas (oxygen/carbon dioxide) exchange. This alone would suffice as a reason for needing to understand the content of this book. However, there is another reason: fluid needs are not obvious! They are often subtle in their presentation and elusive in their mastery. Furthermore, their subtle presentation may not seem so difficult when observed in a vacuum, but in a real life hospital setting, it is rarely so apparent. The initial stress response to blood loss, for example, is something that is shared by many different conditions. Dehydration, pain, inflammation, infection, and other triggers can activate a response similar to that of blood loss in which catecholamines, Ang II, aldosterone, etc. are secreted. Therefore, it is more important to understand the integrated science behind fluid management than to simply memorize what types of fluid are used to treat a particular problem. That said, let's talk for a brief moment about types of fluid replacement.

As we mentioned previously, clinicians can deliver fluids conveniently through two routes: the mouth (PO) and the veins (IV). We touched on oral rehydration therapy a bit earlier, so now we're going to give a bit of a primer on IV fluids. Intravenous fluids were not developed until the mid-nineteenth century. The ultimate goal of IV fluids can be divided into the following:

1. Replace fluid losses
2. Short-term maintenance of circulating blood volume (in order to maintain blood pressure)
3. Deliver basic nutrients (sugar and electrolytes) to a person that is unable to eat or drink.

The reality of bedside fluid management is incredibly more complex than what we're presenting here. In our previous discussions, most of the examples thus far involved extreme situations: massive fluid gain or fluid loss, full blown CHF, or a perfectly healthy subject. Or they've involved us giving them too much or too little fluid replacement. The large majority of patients are not so simple. They might only have mild heart failure, nephrotic syndrome, or liver failure. They may have only moderate vomiting or diarrhea. They are all nevertheless subject to some amount of isotonic, hypotonic, or hypertonic fluid alterations. So keep in mind that there is still an elegance involved in titrating fluids at a patient's bedside!

The classic maintenance of IV fluids (when patients are kept NPO) used in hospitals span a huge amount of fluid types; however, the most common are:

1. 0.9% NaCl solution or normal saline (NS)—which contains 154 mEq of Na^+ and 154 mEq of Cl^- in 1 L of water.

2. 5% dextrose or D5W—which contains 50 g of dextrose (sugar) dissolved in 1 L of water.
3. Lactated ringers solution (LR)—which contains 130 mEq of Na^+, 109 mEq of Cl^-, 4 mEq of K^+, 3 mEq of Ca^{2+}, and 28 mEq of lactate diluted in 1 L of water.

Now, you also have various hypotonic fluids such as D5 ½ NS which is 5% dextrose with 0.45% NaCl solution, fluids with albumin, etc. But to keep it simple, we'll just focus on the first three.

As we mentioned before, these three IV solutions are used to "replace" fluid that is lost through urine, perspiration, GI losses, breathing, sweating, and general evaporation as well as that fluid lost iatrogenically via nasogastric tubes, ileosotomies, and bleeding. It is also used to provide some sort of water, sugar, and electrolyte intake in patients who are not allowed or incapable of eating/drinking in the hospital. These fluids or combinations of them have their indications and particular uses, which vary from patient to patient. However, for the purposes of our discussion, let us clarify some characteristics associated with them.

- NS and LR distribute freely within the EC compartment but DO NOT alter cell size. This is because they are isotonic! They have the same osmolality as serum but their components are relatively impermeable to the cell membrane. So they can move between IV and IT, but not freely between ECF and ICF (remember that the endothelium is permeable to salt and water, while the cell membrane isn't!).
- The varying electrolyte concentration in each solution generates different responses from the body, therefore they must be used with caution. The high Cl^- content in NS can lead to the development hyperchloremia. LR on the other hand can drop a patient's Na^+ concentration (the body has a Na^+ of 140 mEq and LR of 130 mEq). Therefore, all IV fluids, even isotonic ones, must be used with caution!
- WARNING—D5W (50 g of dextrose in 1 L of water), although iso-osmolar, is the equivalent of adding WATER to the body. This means that when we place a patient on D5W, we're effectively diluting them! Why? The two components of D5W are water and sugar. When the water and sugar enter the vascular space, the osmolality is similar to that of plasma, and therefore, initially the RBCs that

come in contact with the D5W don't explode. The dextrose however is rapidly metabolized by the tissues, which effectively only leaves water, which in turn will be distributed in BOTH the IC and the EC compartment because water is permeable to both the endothelium and the cell wall. Therefore, D5W is NOT adequate when replacing lost isotonic fluid!

- Combinations of D5W and NS give us variations in salt and sugar content that can be used to replace the daily requirements of salt, water, and glucose that a patient who is not eating anything needs. The classic teaching is that by adding some glucose to the IV solutions, protein catabolism will be prevented initially, thus preserving a non-eating patient's lean mass. It will prevent also the stress response associated with starvation.

In short, all solutions have benefits and drawbacks and must be used wisely and cautiously, as the unchecked infusion of an isotonic solution (LR and NS) can quickly lead to the development of pulmonary edema, exacerbation of congestive heart symptoms, acidemia, and even renal failure, while the infusion of a hypotonic solution (D5W, or D5 ½ NS) can rapidly give us hypotonic fluid gain with a consequent decrease in ECF osmolality.

7.5 CONCLUSION

The integrated picture of fluid and electrolyte physiology is one of the incredible complexities. Over these past few chapters, we have attempted to provide you with a primer and the basic knowledge that will allow you to move upward in your understanding of these difficult topics. The "broad strokes" approach that we've included in this book obviates many subtleties that will become readily apparent as you continue to study; however, we hope that by presenting the information in a clear "black-and-white" way, the foundation for a successful and enjoyable understanding of fluid electrolyte physiology has been laid.

7.6 CLINICAL VIGNETTES

A 70-year-old man with a history of a myocardial infarction and slight CHF is admitted to the hospital with acute onset intractable nausea and vomiting. He reports having eaten at a food cart in the street with doubtful sanitation. He has vomited approximately 2 L of nonbilious yellow gastric fluid. He mentions that even though he is thirsty, he can't keep anything down.

1. *The loss of gastric fluid (which is characteristically a yellowish color) implies the loss of a large load of H^+ and Cl^- ions. In this particular patient, the kidney's primary goal would be to:*
    A. Retain as much NaCl and water as possible in order to expand the EC volume.
    B. Retain as much HCl as possible in order to correct the acid–base derangement.
    C. Excrete as much HCO_3^- as possible in order to compensate for the loss of HCl through vomiting.

    Answer A. The loss of any kind of isotonic fluid will trigger salt and water, that is, volume retention by the kidney. The primary goal the kidney has is to correct the lost volume. Once the lost volume is recovered, the kidney will then begin to address the acid–base status. Therefore, retention of HCl or excretion of HCO_3^- would not be an appropriate response by the kidney.

2. *After triage and evaluation, the patient is placed on a normal saline or NS (0.9% NaCl in 1 L of water) intravenous drip at 125 mL/h. The patient is admitted to the hospital where he continues to have nausea but no vomiting. After approximately 24 h and 3,000 ml of NS, the patient feels better. However, he notices that his feet and hands are swollen. The administration of NS in this patient had what effect on IC volume:*
    A. Increased IC volume
    B. Decreased IC volume
    C. No change in IC volume

    Answer C. NS is isotonic with plasma. This means that NS will NOT generate fluid shifts between the ECF and ICF because it will not change the ECF osmolality. Through the administration of NS, the volume that was lost through vomiting will be replaced. The problem however is that because NS is composed of water, Na^+, and Cl^-, it will freely permeate from the IV space, where it is being infused into the IT space. As the IT space is approximately 2/3 of the ECF and the IV space is approximately 1/3 of the ECF, 2/3 of the total amount of NS that is administered will end up as IT fluid! This explains why the patient felt that both his hands and feet were swollen after the administration of fluid, since approximately 2 l of the 3 l that where administered over a 24 h period are now in the IT space.

3. *After the administration of 3 L of NS, the patient's urine gets lighter in color (less concentrated) and our patient's thirst decreases (signs that the initial volume loss has been corrected). He begins a diet of clear liquids and then progresses to regular diet. All the while however the NS has been going at a rate of 125 mL/h. Over the course of the next 2 days, the patient's weight goes from 70 to 74 kg. He begins to have respiratory difficulty and progressive hemoglobin desaturation from 97% on room air oxygen to 89%. Treatment of this patient's condition should be aimed at:*
   A. Decreasing IC volume
   B. Decreasing IT volume
   C. Increasing IV volume

Answer B. The patient is suffering from acute pulmonary edema. The excessive administration of normal saline led to excessive IT fluid accumulation. In the lungs, the accumulation of fluid in the IT space thus leading to a decreased diffusion of oxygen into the blood. The treatment for this problem is diminishing IT volume through the administration of a loop diuretic such as furosemide. The administration of furosemide will lead to a loss of Na^+ and water in the urine, therefore eliminating some of the excess ECF.

FURTHER READING

We have chosen to recommend several review articles and books which can aid in further understanding of the basic concepts presented in this book. We have general references, which address most of the concepts in our book, and specific references, which address chapter-specific concepts.

GENERAL REFERENCES

[1] Boron WF, Boulpaep EL. Medical physiology: a cellular and molecular approach. 2nd ed. Philadelphia, PA: Saunders/Elsevier; 2009.

[2] Hall JE, Guyton AC. Guyton and Hall textbook of medical physiology. 12th ed. Philadelphia, PA: Saunders/Elsevier; 2011.

[3] Marino PL, Sutin KM. The ICU book. 3rd ed. Philadelphia, PA: Lippincott Williams & Wilkins; 2007.

[4] Halperin ML, Kamel KS, Fluid MB. Electrolyte and acid-base physiology: a problem-based approach. 4th ed. Philadelphia, PA: Saunders/Elsevier; 2010.

SPECIFIC REFERENCES
Chapter 1

[1] Burg MB. Molecular basis of osmotic regulation. Am J Physiol 1995;268(6 Pt 2):F983–96.

[2] Darrow DC, Yannet H. The changes in the distribution of body water accompanying increase and decrease in extracellular electrolyte. J Clin Invest 1935;14(2):266–75.

Chapter 2

[1] Badeer HS. Hemodynamics for medical students. Adv Physiol Educ 2001;25(1-4):44–52.

[2] Takata M, Wise RA, Robotham JL. Effects of abdominal pressure on venous return: abdominal vascular zone conditions. J Appl Physiol 1990;69(6):1961–72.

[3] Mulvany MJ, Aalkjaer C. Structure and function of small arteries. Physiol Rev 1990;70 (4):921–61.

Chapter 3

[1] Kobayashi T, Solaro RJ. Calcium, thin filaments, and the integrative biology of cardiac contractility. Annu Rev Physiol 2005;67:39–67.

[2] Magder S. The classical Guyton view that mean systemic pressure, right atrial pressure, and venous resistance govern venous return is/is not correct. J Appl Physiol 2006;101(5):1533.

[3] Henderson WR, Griesdale DE, Walley KR, Sheel AW. Clinical review: Guyton—the role of mean circulatory filling pressure and right atrial pressure in controlling cardiac output. Crit Care 2010;14(6):243.

[4] Moss RL, Fitzsimons DP. Frank–Starling relationship: long on importance, short on mechanism. Circ Res 2002;90(1):11–3.

Chapter 4

[1] Wolff CB. Normal cardiac output, oxygen delivery and oxygen extraction. Adv Exp Med Biol 2007;599:169–82.

[2] Hsia CC. Respiratory function of hemoglobin. N Engl J Med 1998;338:239–48.

Chapter 5

[1] Arroyo JP, Ronzaud C, Lagnaz D, Staub O, Gamba G. Aldosterone paradox: differential regulation of ion transport in distal nephron. Physiology 2011;26(2):115–23.

[2] Lote CJ. Principles of renal physiology. 5th ed. New York, NY: Springer; 2012.

[3] Kaplan LJ. It's all in the charge. Crit Care Med 2005;33(3):680–1.

Chapter 6

[1] Arroyo JP, Gamba G. Advances in WNK signaling of salt and potassium metabolism: clinical implications. Am J Nephrol 2012;35(4):379–86.

[2] Magder S. Bench-to-bedside review: an approach to hemodynamic monitoring—Guyton at the bedside. Crit Care 2012;16(5):236.

[3] Gutierrez G, Reines HD, Wulf-Gutierrez ME. Clinical review: hemorrhagic shock. Crit Care 2004;8(5):373–81.

[4] Choi PT, Yip G, Quinonez LG, Cook DJ. Crystalloids vs. colloids in fluid resuscitation: a systematic review. Crit Care Med 1999;27(1):200–10.

Chapter 7

[1] Piper GL, Kaplan LJ. Fluid and electrolyte management for the surgical patient. Surg Clin North Am 2012;92(2):189–205 [vii]

[2] Kaplan LJ, Kellum JA. pH, ions and electrolytes. Curr Opin Crit Care 2010;16(4):323–31.

[3] Adrogue HJ, Madias NE. Hyponatremia. N Engl J Med 2000;342(21):1581–9.

[4] Sam R, Feizi I. Understanding hypernatremia. Am J Nephrol 2012;36(1):97–104.

Printed and bound by CPI Group (UK) Ltd, Croydon, CR0 4YY

03/10/2024

01040421-0013